FROM ER TO E.T.

FROM ER TO E.T.
How Electromagnetic Technologies Are Changing Our Lives

RAJEEV BANSAL

The IEEE Press Series on Electromagnetic Wave Theory
Andreas C. Cangellaris, *Series Editor*

WILEY

For general information on our other products and services or for technical support, please contact our Customer Care Department within the United States at (800) 762-2974, outside the United States at (317) 572-3993 or fax (317) 572-4002.

Wiley also publishes its books in a variety of electronic formats. Some content that appears in print may not be available in electronic formats. For more information about Wiley products, visit our web site at www.wiley.com.

Library of Congress Cataloging-in-Publication Data is available.

ISBN: 978-1-118-45817-4

Printed in the United States of America

10 9 8 7 6 5 4 3 2 1

CONTENTS

ABOUT THE AUTHOR

Rajeev Bansal received his Ph.D. in Applied Physics from Harvard University in 1981. Since then he has taught and conducted research in the area of applied electromagnetics at the University of Connecticut where he is currently a Professor (1992–) and Head (2009-) of the Department of Electrical & Computer Engineering (ECE). His technical contributions include three edited books [*Fundamentals of Engineering Electromagnetics* (2006), *Engineering Electromagnetics: Applications* (2006), *Handbook of Engineering Electromagnetics* (2004)], two commercialized patents (1989 and 1993), and more than 100 journal/conference papers/book chapters. Dr. Bansal has served as an Editor/Reviewer of *Journal of Electromagnetic Waves and Applications* as well as an Associate Editor of *Radio Science*. He has also served on the editorial advisory boards of *IETE Technical Review* and the *International Journal of RF & Microwave CAE*. He is a columnist for the *IEEE Antennas and Propagation Magazine* (1987–) and the *IEEE Microwave Magazine* (since its inception in 2000). He is a fellow of the Electromagnetics Academy, an elected member of the Connecticut Academy of Science and Engineering (CASE), and a senior member of the IEEE. He has served as a consultant to the Naval Undersea Warfare Center, Newport, RI.

PREFACE

Electromagnetic technologies have seeped into every corner of our lives. From that cup of coffee one reheats in the microwave oven to the cellular wireless network that lets one download an ebook in a jiffy, we depend on the marvels of these technologies every day. There are times when we wonder if our privacy is going to be made obsolete one day by RFID chips; others when we worry about the safety of all this radiation around us. Love it or hate it, one cannot get away from these electromagnetic technologies.

In this book, I have tried to showcase many of these wonderful electromagnetic technologies that are changing the world we live in (e.g., new medical devices for the ER) and the future they may create for us (e.g., making contact with E.T. some day). The book owes its genesis to the regular columns for the *IEEE Antennas and Propagation Magazine* and the *IEEE Microwave Magazine* that I have been writing for many years. Based on the comments I have received, the columns have been enjoyed by science/engineering students, practicing engineers, academic colleagues, and many members of the general public with an interest in technology. In planning the book, I had two goals in mind:

1. Preserve the original math-free style of the original columns to make the material accessible to the broadest possible audience and
2. Create a handy textbook *supplement* for students and instructors in courses on electromagnetics (and related fields) by arranging the material in a framework that includes additional technical details and links to electromagnetic textbooks.

With respect to the second goal, it may be noted in passing that the current accreditation criteria for electrical engineering programs emphasize that students, in addition to mastering the technical content, become conversant with the *societal* and *ethical* implications of technologies and learn to place the engineering subject matter in the broader *global* context. This small book goes some distance in filling that niche in a student's technical education. Depending on the topic, the tone of each essay varies but, in my opinion, the material is presented always in a readily accessible and succinct style so that it can be read quickly in a classroom and discussed without having to take too much time away from the technical material being covered in the class. (I have used this approach with some of these essays myself in courses ranging from a freshmen class open to non-engineers to junior/senior level EM/microwaves courses.) Another goal is to entice the reader into pursuing the sources for the essay to delve more deeply into the subject (*life-long learning*). The sources are always listed in each short piece and most of them are available online for instant access.

The main technical entries in the book are grouped by broad areas of application/interest and, in the spirit of Monty Python's "And Now for Something Completely Different," are interspersed with amusing tidbits in the form of "quizzes" and essays on far-out topics. I hope you will have as much fun reading them as I had in compiling them.

ACKNOWLEDGMENTS

I would like to thank the editors-in-chief Ross Stone (through 2014) and Mahta Moghaddam of the *IEEE Antennas and Propagation Magazine* (where most of the columns originally appeared) and the editors of the *IEEE Microwave Magazine* for their support over the years. I am also thankful to Taisuke Soda, the then acquisitions editor for Wiley/IEEE, as well as the reviewers of the book proposal and a draft of the manuscript for their many helpful suggestions. I appreciate greatly the support of the current Wiley/IEEE editor Mary Hatcher, who kindly extended the deadline to complete this book when my additional short-term professional responsibilities forced a 2-year delay in the project. She also helped me find an online resource (Pixabay) for the images used in this book. Finally, I would like to express my deep gratitude to my family, without whose encouragement the book would have remained a gleam in the eye.

Rajeev Bansal

CHAPTER 1

ON THE SHOULDERS OF GIANTS

"If I have seen further it is by standing on the shoulders of giants."
—Isaac Newton (1642–1727)

1.1 HE(A)DY STUFF

Hedy Lamarr (1914–2000), a celebrated movie actress from the Golden Age of Hollywood, once said, "Any girl can be glamorous. All she has to do is stand still and look stupid." Well, the very same Lamarr, in her spare time, co-invented—yes, I'm not kidding—a frequency-hopping radio-controlled system for guiding torpedoes. For

From ER to E.T.: How Electromagnetic Technologies Are Changing Our Lives, First Edition. Rajeev Bansal.
© 2017 by The Institute of Electrical and Electronic Engineers, Inc. Published 2017 by John Wiley & Sons, Inc.

her invention, US Patent #2,292,387, granted in 1942, Hedy Lamarr finally received some long-overdue recognition. On March 12, 1997, at a San Francisco ceremony, her son accepted the Electronic Frontier Foundation Award, given to "Hedy Lamarr for her Contribution in Pioneering Electronics."

This bizarre sequence of events had its roots in pre-war Vienna, where, in 1933, the Austrian-born Hedy Lamarr married Fritz Mandl, a leading Austrian armaments manufacturer. Mandl's household was an institution in the Viennese society, attracting many dignitaries, including political and military leaders. Mandl was himself interested in control systems, and engaged in research in that field. Apparently, Hedy Lamarr also picked up a thing or two along the way. In 1937, their marriage broke up, and Lamarr emigrated to America and headed for Hollywood.

One day, in the summer of 1940, Hedy Lamarr and her Hollywood neighbor, George Antheil, an avant-garde composer, were playing the piano together, carrying on an improvised musical dialogue up and down the keyboard. They started talking about the war and Lamarr brought up the idea of frequency hopping, synchronized between the transmitter and the receiver, for secure (resistant to jamming) radio control of torpedoes. Antheil, drawing upon his experience with player pianos, suggested that the synchronized rapid frequency hopping that Lamarr had envisioned for the torpedo-control system could be implemented using perforated paper rolls, similar to player-piano rolls. In fact, by the time they applied for a patent in June 1941, their embodiment of the frequency-hopping technique used slotted paper rolls and utilized 88 frequencies, the exact number of keys on a piano. Their patent application also specified that the torpedo could be guided from above by a plane.

While their invention was granted a patent in 1942, it was an entirely different ball game when it came to convincing the US Navy that the device was practical for torpedo control. Ironically, Antheil's contribution of the player-piano mechanism as *one* possible implementation of Lamarr's frequency-hopping system apparently proved to be its undoing. Antheil later wrote, "In our patent Hedy and I attempted to better elucidate our mechanism by explaining that certain parts of it worked like the fundamental mechanism of a player piano. Here, undoubtedly, we made our mistake. The reverend and brass-headed gentlemen in Washington who examined our invention read no further than the words "player piano." "My God," I can see them saying, "we shall put a player piano in a torpedo." Amusing as Antheil's explanation is, it probably does not tell the whole story.

The Navy must have also realized the difficulties in setting up an electromagnetic communication link between a plane and a torpedo through the highly attenuating sea water (a problem that persists to this day) and, in particular, in equipping the torpedo with a suitable receiving antenna.

At any rate, discouraged by the Navy's attitude, Lamarr and Antheil did not pursue their invention further. Instead, Lamarr successfully used her charm to sell war bonds, raising millions of dollars for the war effort. In 1957, engineers at Sylvania Electronic Systems Division in Buffalo, New York, implemented secure radio communication via frequency hopping by replacing the piano rolls with electronic circuits. In 1962, 3 years after the Lamarr–Antheil patent had expired, ships equipped with secure military-communication systems, based on the frequency-hopping technique, were deployed during the Cuban missile crisis. Though Lamarr and Antheil never collected a penny from their pioneering work, their patent has been cited as the seminal work by subsequent patents in the area of frequency-hopping systems.

The bottom line is that Lamarr and Antheil were inventors ahead of their time. That is why it is particularly fitting that the 1997 Electronic Frontier Foundation Award finally recognized the vital role their patent eventually played in the modem development of secure communications.

[Compiled from reports in the *Chicago Tribune,* March 31, 1997, and the *American Heritage of Invention & Technology*, Spring 1997.]

(The original version of the column appeared in "AP-S turnstile," *IEEE Antennas and Propagation Magazine*, vol. 39, no. 3, p. 100, June 1997.)

NOTES

1. Many books are available about Hedy Lamarr and her inventions. See, for example, Richard Rhodes's *Hedy's Folly: The Life and Breakthrough Inventions of Hedy Lamarr, the Most Beautiful Woman in the World*, Vintage, 2012.
2. For more information about US Patent #2,292,387, consult: http://www.google.com/patents/US2292387 (accessed December 22, 2015).

1.2 FROM RUSSIA WITHOUT ENGLISH

The National Academy of Sciences recently published a biographical memoir [1] of Edward Leonard Ginzton (1915–1998). Ginzton's fundamental contributions to electronics and microwaves have been eulogized before, for example, when he was elected to the National Academy of Engineering (1965), or when he received the IEEE Medal of honor (1969). What Anthony Siegman, the author of this brief 35 page memoir, succeeds in doing is to present a multi-faceted portrait of Ginzton, who in Carolyn Caddes's words was "scientist, educator, business executive, environmentalist, and humanitarian [2]." Here are some highlights from Siegman's memoir.

Early Years

Ginzton, who was born in Ukraine, recalls his childhood education thus: "Since both of my parents participated as medical officers on the Eastern Front, my early childhood consisted of rapid migration with the tides of war, revolution, and other similar events. Until I was 8 we did not live in any one place for more than six months, and I was not exposed to formal education until I was 11." When Ginzton was 13, his family emigrated from Russian-controlled Manchuria to San Francisco. Ginzton, "knowing not a word of English," was placed in the first grade in the public schools. Four years later, Ginzton finished high school and entered UC Berkeley.

The War Years

As Ginzton later noted modestly: "[During these] six years, I invented some 40 or 50 devices, some of which were relatively important." The Doppler radar techniques pioneered under Ginzton's supervision at Sperry provided the technical foundation of many sophisticated radars of later years. By the time the war ended, Ginzton had some 2000 people working under his direction.

The Stanford Years

Appointed an assistant professor of applied physics (rather than physics because of his EE credentials) at Stanford, Ginzton and his colleagues published "A Linear Electron Accelerator" in the February 1948 issue of the *Review of Scientific Instruments*, which led to the successful development of several generations of accelerators, the use of which was instrumental in at least six Nobel Prizes. Ginzton was also prescient in recognizing the potential application of these accelerators for radiation therapy for cancer; by the time of his death some 4000 of small medical accelerators were treating over 1 million patients annually.

The Varian Years

Varian Associates was established in 1948 with $22,000 of capital and six full-time employees. Ginzton was on the company's board from day one and remained on it till 1993. In 1959, when Russell Varian died suddenly, Ginzton was appointed CEO.

The McCarthy Era

Ginzton, who had shared a graduate-student office with Frank Oppenheimer (Robert's brother) in 1939, was caught up in the McCarthy maelstrom and lost his security clearance for a while. It required a massive legal effort by Stanford to get Ginzton's clearance back.

Community Leadership

Ginzton was an early champion of fair housing and clean air. He also founded and co-chaired (1968–1972) the Stanford Mid-Peninsula Urban Coalition, an organization to support the launch of minority-owned small businesses.

His Vision

Writing in 1956, Ginzton noted: "It is evident that the applications of present microwave knowledge will continue to grow, both in number and diversity; but despite the daily invention of novel applications, we must not become complacent. Every field of research has a finite half-life." Later in his life, he explained his philosophy to Caddes [2]: "Grow and become educated, but do not equate professional training with education. Try to learn how to think. Attempt to do what you want to do. Making a living is not enough."

ACKNOWLEDGMENT

I would like to thank my colleague Dr. Anthony DeMaria for bringing the Ginzton memoir to my attention.

REFERENCES

[1] A. E. Siegman, "Edward Leonard Ginzton (1915–1998)," *Biographical Memoirs*, vol. 88, National Academy of Sciences, Washington, D.C., 2006.
[2] C. Caddes, *Portraits of Success: Impressions of Silicon Valley Pioneers*, Tioga, Palo Alto, CA, 1986.

(The original version of the column appeared in "Microwave surfing," *IEEE Microwave Magazine*, vol. 7, no. 6, pp. 28–30, December 2006.)

1.3 ON THE SHOULDERS OF GIANTS

We know Maxwell's equations by heart but what do we know about Maxwell himself? Considering his preeminent position in the pantheon of leading scientists, there are relatively few biographies of Maxwell. The primary historical reference on Maxwell has been the 1882 biography written by Maxwell's long-time friend Lewis Campbell with help from William Garnett. Campbell's work received critical acclaim upon publication. "This volume will be heartily welcomed by all who knew Clerk Maxwell, and who cherish his memory, and lay the still wider circle of those who derive pleasure and new vigour from the study of the lives and work of the great men that have gone before them," noted the reviewer in *Nature* but it is no longer readily available. Fortunately, the full text of the 1882 biography is available on the web at https://www.sonnetsoftware.com/resources/maxwell-bio.html

Campbell's book, *The Life of James Clerk Maxwell*, has three parts: (1) a description of Maxwell's life, (2) an account of his scientific investigations, and (yes, I am not kidding) (3) a collection of Maxwell's poetry. Maxwell's poems cover the full spectrum from translations of Virgil's poetry to original poems on scientific issues. An example of the latter follows (I have omitted the middle section of the rather long poem as well as the accompanying figure and equations, which may be viewed at the website mentioned above):

A PROBLEM IN DYNAMICS (1854)

AN inextensible heavy chain

Lies on a smooth horizontal plane,

An impulsive force is applied at A,

Required the initial motion of K.

Let ds be the infinitesimal link,

Of which for the present we've only to think;

Let T be the tension, and T + dT

The same for the end that is nearest to B.

Let a be put, by a common convention,

For the angle at M 'twixt OX and the tension;

Let Vt and Vn be ds's velocities,

Of which Vt along and Vn across it is;

Then Vn/Vt the tangent will equal,

Of the angle of starting worked out in the sequel.

In working the problem the first thing of course is

To equate the impressed and effectual forces.

K is tugged by two tensions, whose difference dT

(1) Must equal the element's mass into Vt.

Vn must be due to the force perpendicular

To ds's direction, which shows the particular

Advantage of using da to serve at your

Pleasure to estimate ds's curvature.

For Vn into mass of a unit of chain

(2) Must equal the curvature into the strain.

Thus managing cause and effect to discriminate,

The student must fruitlessly try to eliminate,

And painfully learn, that in order to do it, he

Must find the Equation of Continuity.

…

From these two conditions we get three equations,

Which serve to determine the proper relations

Between the first impulse and each coefficient

In the form for the tension, and this is sufficient

To work out the problem, and then, if you choose,

You may turn it and twist it the Dons to amuse.

In 1884, an abridged second edition of the biography was published which included several previously unpublished scientific letters. I particularly enjoyed the following excerpt from a letter that Maxwell sent to Faraday in 1859:

> *"DEAR SIR—I am a candidate for the Chair of Natural Philosophy in the University of Edinburgh, which will soon be vacant by the appointment of Professor J. D. Forbes to St. Andrews. If you should be able, from your knowledge of the attention which I have paid to science, to recommend me to the notice of the Curators, it would be greatly in my favour, and I should be much indebted to you for such a certificate."*

I don't know whether Faraday obliged with a glowing recommendation, but Maxwell didn't get the job!

(The original version of the column appeared in "AP-S turnstile," *IEEE Antennas and Propagation Magazine*, vol. 41, no. 1, p. 116, February 1999.)

NOTE

1. For additional material on Maxwell and his famous equations, see Sections 4.1, 10.4, and 10.5.

1.4   DO-IT-YOURSELF EXECUTION?

An article by Niels Jonassen of the Technical University of Denmark (*Compliance Engineering*, January/February 1998) sparked my interest in Benjamin Franklin's technical writings. In July 1750, Franklin proposed the following experiment in a letter to his British friend P. Collison in London:

> *"To determine the question whether the clouds that contain lightning are electrified or not I would propose an experiment to be tried where it may be done conveniently. On top of some high tower or steeple place a kind of sentry box...big enough to contain a man and an electrical stand [an insulated platform]. From the middle of the stand let an iron rod rise and pass bending out of the door, and then upright twenty or thirty feet, pointed very sharp at the end. If the electrical stand be kept clean and dry, a man standing on it when clouds are passing low might be electrified and afford sparks, the rod drawing fire to him from the cloud."*

An astonishing suggestion, indeed, from the inventor of the lightning rod, particularly when one considers the next statement in Benjamin Franklin's letter: *"If any danger to the man be apprehended (though I think there would be none)..."* Fortunately for Franklin, there was no suitable tower in Philadelphia, so he did not get a chance to try this "glow in the dark" experiment himself! However, his letter received a wide and enthusiastic audience in Europe. In May 1752, the French scientist d'Alibart carried out the experiment near Versailles and lived to tell the tale at the Academy of Sciences in Paris 3 days later. Soon after, the experiment was successfully duplicated in France again, in England, and in Belgium. Next year the Swedish physicist Georg Richman, working in Russia, installed "lightning chords" through the roof of his house with the chords ending above his desk so that he could observe the lightning phenomenon from the comfort of his chair. On July 26, 1753, the top of the lightning chords received a direct lightning strike and, in the memorable words of his colleague Lomonosov, Richman died a splendid death fulfilling a duty of his profession.

Meanwhile, back home in Philadelphia, Franklin, apparently unaware of the European experiments, "improved" on his original idea and thought of the famous kite experiment (which dispensed with the need for a tower). Sometime during the summer of 1752, the classic experiment was performed: sparks jumped from the metal key at the end of the electrified conducting kite string to Franklin's hand. No harm done! Since then a number of people have been killed imitating Benjamin Franklin. In the late nineteenth and the early twentieth centuries, large box kites carrying meteorological instruments were used by many US Weather Bureau stations. The kites used weighed 8 lb, carried a couple of pounds of instrumentation, and dragged a good deal of piano wire ("kite string") behind them. In one case, a man assisting in the flight was killed when the kite was struck by lightning [1].

REFERENCE

[1] M. Uman, *All About Lightning*, Dover, 1986.

(The original version of the column appeared in "AP-S turnstile," *IEEE Antennas and Propagation Magazine*, vol. 40, no. 2, p. 102, April 1998.)

NOTES

1. A good biography of Ben Franklin is: W. Issacson, *Benjamin Franklin: An American Life*, Simon & Schuster, 2003.

2. A farmer and his cow are caught outdoors in a thunderstorm. A pine tree near them is struck by lightning. The cow is electrocuted but the farmer survives to tell the tale. How?
    (a) The cow presents a much larger capacitance than the farmer.
    (b) The cow happens to be a bit closer to the tree.
    (c) The cow's legs are too far apart.
    (d) It is a totally random occurrence.

    (c) The cow's legs are too far apart.
    The current from the lightning strike, which can be tens of thousands of amperes, passes into the earth at the base of the tree and spreads out radially in the top conducting layer of the ground. This sets up a potential gradient along the surface. The cow's foot near the tree will be at a much higher potential (depending on the ground resistance) than the foot farthest away from the tree. Clearly, a current will flow through the cow from one end to the other and could well be fatal.
    Source: R. Bansal, "Zapped: A pop quiz on electrostatics," *IEEE Potentials*, pp. 5–6, April/May 2000.

3. Textbook resources:
    (i) W. H. Hayt and J. A. Buck, *Engineering Electromagnetics*, 8th ed., McGraw-Hill, New York, 2012. Electrostatic fields are discussed in Chapters 2–6.
    (ii) F. T. Ulaby and U. Ravaioli, *Fundamentals of Applied Electromagnetics*, 7th ed., Prentice Hall, Upper Saddle River, NJ, 2015. Electrostatic fields are discussed in Chapter 4.

1.5 FRANKLIN: DID HE OR DIDN'T HE?

Benjamin Franklin's investigations [1] into electricity were discussed previously in Section 1.4. As all American children learn in school, Mr. Franklin showed that lightning was a form of electricity through his celebrated "kite experiment" sometime in 1752. Now a new book [2] by Tom Tucker poses the iconoclastic question: Did Benjamin Franklin *really* fly that kite?

To backtrack a little bit, in July 1750, Franklin had proposed the following experiment in a letter to his British friend P. Collison in London [1, 3]:

> "*To determine the question whether the clouds that contain lightning are electrified or not I would propose an experiment to be tried where it may be done conveniently. On top of some high tower or steeple place a kind of sentry box...big enough to contain a man and an electrical stand [an insulated platform]. From the middle of the stand let an iron rod rise and pass bending out of the door, and then upright twenty or thirty feet, pointed very sharp at the end. If the electrical stand be kept clean and dry, a man standing on it when clouds are passing low might be electrified and afford sparks, the rod drawing fire to him from the cloud.*"

As pointed out in Section 1.4, fortunately for Franklin, there was no suitable tower in Philadelphia, so he did not get a chance to try this dangerous experiment *himself*! However, thanks to his letter, the experiment was carried out successfully several times in Europe. Meanwhile, back home in America, Franklin, apparently unaware of the European demonstrations, refined his original idea and thought of the famous kite experiment, which did not need a tower. Sometime during the summer of 1752, the classic experiment was performed: sparks jumped from the metal key at the end of the electrified conducting kite string to Franklin's hand. Or at least that is the way the story is conventionally told. Tucker, the author of *Bolt of Fate* [2], has his doubts though. He asks us to read carefully the account of the kite experiment Franklin published in the *Pennsylvania Gazette* [4]:

> "*The kite **is to be raised**, when a thundergust appears to be coming on, (which is very frequent in this country) and the person, who holds the string, must stand within a door,*

*or window, or under some cover, so that the silk riband may not be wet... As soon as any of the thunder-clouds come over the kite, the pointed wire **will draw** the electric fire from them; and the kite, with all the twine, **will be electrified**; and the loose filaments of the twine **will stand out** every way, and be attracted by the approaching finger. When the rain has wet the kite and twine, so that it can conduct the electric fire freely, **you will find** it stream out plentifully from the key on the approach of your knuckle... All the other electrical experiments [can] be performed, which are usually done by the help of a rubbed glass globe or tube, and thereby the sameness of the electric matter with that of lightning completely demonstrated" [emphasis added].*

Tucker was intrigued by the conditional spirit and the unusual future tenses used by Franklin in describing this experiment. He compared it with the reports of other experiments conducted by Franklin and found that the impersonal future-tense style used in the excerpt above was typical neither of the eighteenth century scientific reports in general nor of Franklin's own work in particular. In fact, according to Tucker, Franklin was usually very careful in describing exactly how, when, and where he did a particular experiment: "he gives specifics, he uses active voice, he offers diagrams, *he says he did it.*"

In another recent biography [5] of Benjamin Franklin, Issacson does not accept Tucker's analysis. He cites the great historian of science I. Bernard Cohen's books on Franklin's science. Cohen notes that the "kite experiment" was later reproduced by others and it was unlike Franklin to have just made up the account. Tucker, for his part, draws attention to a series of (nonscientific) hoaxes that Franklin pulled in his publishing career.

I would like to conclude this section by quoting from Gopnik [4], who points out that it was Franklin who edited Jefferson's draft to come up with the words "We hold these truths to be *self-evident*":

"The moral of the kite [story] is not that truth is relative. It is that nothing is self-evident.... We hold these truths as we hold the twine, believing, without being sure, that the tugs and shocks are what we think they are. We hold the string, and hope for the best. Often, there is no lightning. Sometimes, there is no kite."

REFERENCES

[1] R. Bansal, "AP-S turnstile: do-it-yourself electrocution?" *IEEE Antennas and Propagation Magazine*, vol. 40, no. 2, p. 102, April 1998.

[2] T. Tucker, *Bolt of Fate: Benjamin Franklin and His Electric Kite Hoax*" PublicAffairs, 2003.

[3] M. Uman, *All About Lightning*, Dover, 1986.

[4] A. Gopnik, "American Electric," *The New Yorker*, pp. 96–100, June 30, 2003.

[5] W. Issacson, *Benjamin Franklin: An American Life*, Simon & Schuster, 2003.

(The original version of the column appeared in "AP-S turnstile," *IEEE Antennas and Propagation Magazine*, vol. 45, no. 4, pp. 82–83, August 2003.)

NOTES

1. 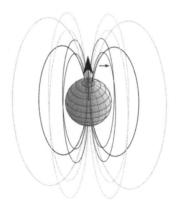 The electric charge delivered to the earth during a lightning strike is around

   (a) 1 nC

   (b) 1 C

   (c) 10^6 C

   (d) 10^{12} C

   > **(b) 1 C**
   >
   > Since the electric filed is defined as the force exerted on a unit charge (1 C in the meter-kilogram-second, MKS, metric system), it is a common mistake to think that 1 C represents a tiny charge. Actually, the MKS unit of charge (named after Charles Coulomb) is way too large for most electrical engineering applications. Recall that a negative charge of 1 C will require almost 6×10^{18} electrons, hardly a "point charge" but rather the type of charge transfer that takes place during a violent phenomenon like a lightning strike.
   >
   > *Source*: R. Bansal, "Zapped: A pop quiz on electrostatics," *IEEE Potentials*, pp. 5–6, April/May 2000.

2. Textbook resources:

   (i) W. H. Hayt and J. A. Buck, *Engineering Electromagnetics*, 8th ed., McGraw-Hill, New York, 2012. Electrostatic fields are discussed in Chapters 2–6.

   (ii) F. T. Ulaby and U. Ravaioli, *Fundamentals of Applied Electromagnetics*, 7th ed., Prentice Hall, Upper Saddle River, NJ, 2015. Electrostatic fields are discussed in Chapter 4.

3. Reference [1] is included in this book as Section 1.4.

1.6 DE MAGNETE ("ON THE MAGNET")

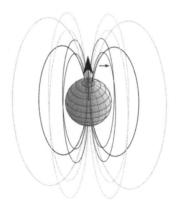

Navigators all at sea

Don't eat onions for their tea

Not that they're at all emetic

They make the compass nonmagnetic

At the dawn of the twenty-first century, it may seem quaint that British naval helms-men were once flogged if they were found to be in violation of the regulation [1] that "steersmen, and such as tend the Mariner's Card are forbidden to eat Onyons and Gar-lick, lest they make the index of the poles drunk." But such was indeed the sixteenth century world into which William Gilbert (1544–1603) was born. The year 2000 marked the 400th anniversary [2] of the publication of his pioneering treatise "De Magnete" (ix+246 pp, 7/6 (37.5 p) in London; 2 Thaler in Frankfurt; Published by Peter Short, London, 1600), which eventually helped dispel many nonsensical beliefs about magnetism. Predating Kepler's *Astronomia Nova* (1609), which described laws of planetary motion, and Galileo's *Sidereus Nuncius* (1610), which reported on his astronomical observations with the telescope, Gilbert's book on magnetism is considered by many to be the first scientific monograph written on modern principles. (Newton's *Principia* came much later in 1687.) Like a present-day doctoral disser-tation, "De Magnete" [3] reviews previous work, describes Gilbert's experimental investigations and results, discusses his findings in a broader context, and finally provides speculations about future work [4].

William Gilbert was born in 1544 in Colchester (some 50 miles NE of London), where his father held the prestigious post of Recorder. Gilbert studied at St John's College, Cambridge, where he remained for 11 years until 1569, acquiring bachelor's and master's degrees, as well as his credentials as a physician. He settled in London to practice medicine and eventually became President of the Royal College of Physi-cians and Physician to Queen Elizabeth I. He died of bubonic plague in London in 1603 [4].

Gilbert's research on magnetism was really a hobby. Over a 20 year period (1581–1600), he conducted experiments in electrostatics (contributing to our vocabulary terms such as "electrick force") and in magnetostatics. For his work on geomagnetism, Gilbert constructed a spherical lodestone which he called terrella ("little earth"). Using small compass needles, he explored the magnetic field of his terrella and extrapolated his findings to the effect of the earth's magnetic field on the behavior of a magnetic compass. For example, to simulate the effects of proximity to landmasses, he carved out "oceans" from the terrella and found that the compass needles behaved differently near oceans and mountain ranges [4].

Gilbert supported the Copernican theory and also believed that the earth turned on its axis. However, he erroneously associated this planetary rotation with magnetism. The concept of a revolving earth was so heretical at the time that the continental copies of his book had the related pages expunged [4].

To celebrate Gilbert's contributions to magnetism, the American Geophysical Union, which publishes *Radio Science*, held a special session at its Spring 2000 meeting in Washington, D.C. The AGU journal *Eos* also published a belated (!) "book review" of "De Magnete," which is accessible online with a lot of other fascinating details about Gilbert's work and geomagnetism on a NASA website [4] maintained by Dr. David Stern.

If you fast forward the history of magnetism to the year 2000, you may wish to take note of some recent work on **nanomagnets** reported in the *Journal of Applied Physics* [5]. A cooled mixture of iron oxide, polystyrene, and methanol under the influence of a strong magnetic field behaves like a jarful of tiny compasses (nanomagnets). The Barcelona group, which did the work on nanomagnets, expects that they may someday lead to super-fast electronic switches and more efficient cores for power equipment. However, in making predictions about how soon nanomagnets may find real-life applications, it may be worthwhile to remember that it was almost 100 years after the publication of "De Magnete" before the flogging of British helmsmen with garlic-breath finally stopped.

REFERENCES

[1] B. Bolton, *Electromagnetism and its Applications*, VNR, New York, 1980.

[2] W. Leary, "Celebrating the Book That Ushered In the Age of Science", *The New York Times* [Online]. Available: http://www.nytimes.com/2000/06/13/science/celebrating-the-book-that-ushered-in-the-age-of-science.html (accessed March 17, 2016).

[3] W. Gilbert, *De Magnete* (A reprint of Mottelay's 1893 English translation), Dover, New York, 1991.

[4] D. Stern, "The Great Magnet, the Earth" [Online]. Available: http://www-spof.gsfc.nasa.gov/earthmag/demagint.htm (accessed March 17, 2016).

[5] "Nanomagnets: Lodestones on the Loose," *The Economist* [Online]. Available: http://www.economist.com/node/348579 (accessed (March 17, 2016).

(The original version of the column appeared in "AP-S turnstile," *IEEE Antennas and Propagation Magazine*, vol. 42, no. 5, pp. 110–111, October 2000.)

NOTES

1. Here is a very brief chronology of the *other* key figures in the history of electromagnetism:
   (i)   Charles-Augustin de **Coulomb** (1736–1806): Coulomb's law for the electrostatic force between charges
   (ii) Andre-Marie **Ampere** (1775–1836): Ampere's law for the force between current-carrying wires
   (iii) Michael **Faraday** (1791–1867): Faraday's law of electromagnetic induction
   (iv) James Clerk **Maxwell** (1831–1879): Maxwell's equations for electromagnetism

2. Which produces the *largest* magnetic field?
   (a) An MRI (magnetic resonance imaging machine)
   (b) The earth
   (c) A 13 kV distribution line along the street
   (d) A 735 kV transmission line from a power generating station

   (a) An MRI (magnetic resonance imaging machine)
   MRI machines use very large static magnetic fields (more than ten thousand gauss) for diagnostic purposes. Among the remaining choices, the earth's *static* (dc) magnetic flux density in air (about half a gauss) is roughly 200 times larger than the ac magnetic

flux density from typical 60 Hz distribution lines. Lastly, since the magnetic field varies with the line *current,* a 13 kV distribution line may (in some situations) generate a larger magnetic field than a 735 kV transmission line [1 T (tesla) = 1 Wb/m^2 = 10,000 G (gauss)].

Source: R. Bansal, "Pop quiz: EMF and your health," *IEEE Potentials*, pp. 3–4, August/September 1997.

3. Textbook resources:

  (i) W. H. Hayt and J. A. Buck, *Engineering Electromagnetics*, 8th ed., McGraw-Hill, New York, 2012. Magnetostatic fields are discussed in Chapter 7.

  (ii) F. T. Ulaby and U. Ravaioli, *Fundamentals of Applied Electromagnetics*, 7th ed., Prentice Hall, Upper Saddle River, NJ, 2015. Magnetostatic fields are discussed in Chapter 5.

1.7 A EUREKA MOMENT

Ten years ago, while I was in China accompanying my younger daughter on a high-school "field trip," I had the opportunity to experience the magnetic levitation technology firsthand when our group boarded the Shanghai Maglev Train [1] on the way to the Pudong International airport. Although the ride lasted only around 7 minutes, the 267mph top speed certainly provided an exhilarating experience. While many maglev transportation projects have been proposed over the years (e.g., a maglev link between Las Vegas and Disneyland [2]) and some have even reached the demonstration stage in the United States, for most people magnetic levitation remains a scientific curiosity, good enough for wowing students in a freshman physics laboratory with an inexpensive experimental kit such as [3], but with few realizable practical applications. More recently, a Harvard research group funded by the Bill and Melinda Gates Foundation came up with an ingenious way [4] to use the maglev technology for the on-the-go testing of the purity of food and water. And, amazingly, the device was expected to cost less than $50 (a big plus in the developing countries), even less than the price of a simple maglev demo kit such as the one from PASCO [3].

Everybody has heard about Archimedes's (287–212 BC) Eureka moment. The ancient Greek polymath had been asked by the king of Syracuse to determine whether a certain crown was made of pure gold. Archimedes realized that the density of the crown could be used as a simple test of purity. While the weight was easy to measure, the question was how to determine the volume of an irregularly shaped object. The story [5] is told that as Archimedes lowered himself into a bath he noticed that some of the water was displaced by his body and flowed over the edge of the tub. This was just the insight he needed to realize that the crown should not only weigh the right amount but should displace the same volume as an equal weight of pure gold. He was so excited by this idea that he reportedly ran naked through the streets shouting "Eureka" ("I have found it").

Fast forward to the twenty-first century and one finds that density measurement is still a handy way for determining the composition of many substances such as water (how much salt is there in irrigation water?) and milk (what is the fat content?). While sophisticated techniques exist for measuring the density, the required equipment is either too expensive or not portable enough. Professor George Whitesides's research group at Harvard described in a paper [6] published in the *Journal of Agricultural and Food Chemistry* "*a method and a sensor that use magnetic levitation (MagLev) to characterize samples of food and water on the basis of measurements of density. The sensor comprises two permanent NdFeB magnets positioned on top of each other in a configuration with like poles facing* [so that the magnetic field has its minimum at the center of the gap] *and a container filled with a solution of paramagnetic ions. Measurements of density are obtained by suspending a diamagnetic object* [e.g., a drop of milk or water] *in the container filled with the paramagnetic fluid, placing the container between the magnets, and measuring the vertical position of the suspended object* [with a vertically mounted ruler]. [A diamagnetic object will be pushed toward the center of the gap by the magnetic forces and below it by gravity, coming to an equilibrium point dependent on its density.] *MagLev was used to estimate the salinity of water, to compare a variety of vegetable oils on the basis of the ratio of polyunsaturated fat to monounsaturated fat, to compare the contents of fat in milk, cheese, and peanut butter, and to determine the density of grains.*" Archimedes would have been pleased.

REFERENCES

[1] Shanghai Maglev Train website [Online]. Available: http://www.smtdc.com/en/ (accessed March 17, 2016).

[2] E. Werner and K. Hennessey, 300-mph train would whisk travelers from Vegas to Disneyland, *USA Today* [Online]. Available: http://www.usatoday.com/travel/news/2008-02-25-vegas-disneyland-train_N.htm# (accessed March 17, 2016).

[3] Magnetic Levitation demonstration kit in the PASCO engineering catalog (2011).

[4] "Fast-Track Testing," *The Economist Technology Quarterly*, pp. 10–12, September 4, 2010.

[5] A webpage devoted to Archimedes [Online]. Available: http://www.ancientgreece.com/s/People/Archimedes/ (accessed March 17, 2016).

[6] K. Mirica, S. Phillips, C. mace, and G. Whitesides, "Magnetic levitation in the analysis of foods and water," *Journal of Agricultural and Food Chemistry*, vol. 58, pp. 6565–6569, May 2010. Available: http://dx.doi.org/10.1021/jf100377n.

(The original version of the column appeared in "AP-S turnstile," *IEEE Antennas and Propagation Magazine*, vol. 52, no. 3, pp. 174–175, June 2011.)

NOTE

1. Textbook resources:
    (i) W. H. Hayt and J. A. Buck, *Engineering Electromagnetics*, 8th ed., McGraw-Hill, New York, 2012. Magnetic forces are discussed in Chapter 8.
    (ii) F. T. Ulaby and U. Ravaioli, *Fundamentals of Applied Electromagnetics*, 7th ed., Prentice Hall, Upper Saddle River, NJ, 2015. Magnetic forces are discussed in Chapter 5.

1.8 AULD LANG SYNE

"'Technological innovation' is more than just invention. It is a process, often long and costly, of transforming new scientific knowledge into feasible technology, introducing it to use, and making its benefits available to the public. 'Technical integration' is intended to emphasize the more subtle flow of an intangible—engineering information and understanding. Not only has Bell Labs innovated, but it also showed the world technical integration of the innovations. The Bell System Technical Journal was a key enabler for this achievement" [1].

Rod Alferness
Chief Scientist, Bell Labs

In July 1922, when the Information Department of the American Telephone and Telegraph Company printed no. 1, vol. 1 of the now-legendary *The Bell System Technical Journal* (50c. per copy; $1.50 per year), the field of electrical communication was still in its infancy. Most industries relied upon cut-and-try, rule-of-thumb methods for engineering development and the editorial board of the new journal, which included Edwin H. Colpitts of the Colpitts-oscillator fame, could only hope that "the time will undoubtedly come when every industry will recognize the aid it can derive from scientific research in some form as it now recognizes its dependence for motive power on steam or electricity rather than on muscular activity." The field of electrical communication was certainly leading the way in that direction and the editorial board proudly noted: "Electrical communication is credited with having organized a research laboratory prior to the first university course in electrical engineering" [2].

From July 1922 to December 1983 (when it ceased publication), *The Bell System Technical Journal (BSTJ)* published not only original research in the field of electrical communication but also reprints on [1] "important research and development work in the communication field generally so that the results of such work may be given greater publicity and become of greater value to communication engineers." The first issue included papers on high-power vacuum tubes, submarine cables (not fiber optic!), and the analysis of speech signals among others. The last issue (no. 10, vol. 62, December 1983) featured a special section devoted to the engineering behind single-sideband (SSB) microwave radio systems and, on the theoretical side, included a paper on an algebraic theory of relational databases [3]. But, it is the stuff that came in between those bookends that literally makes our communication world go round.

In cooperation with the IEEE, Alcatel-Lucent has recently made the entire *BSTJ* archive (1922–1983) available online [1] to the global research community. Yes, it has a "search" button and, yes, you can download pdf copies of the papers you would like to add to your own personal digital library. So whether you are looking for Shannon's seminal paper on the mathematical theory of communication or trying to trace the development of the cellular technology, you have just been handed the key to the candy store. Step right in and taste as many treats as you like. Bon appetit!

ACKNOWLEDGMENT

I would like to thank my colleague Dr. Anthony DeMaria, who first drew my attention to the online archives (1922–1983) of the *BSTJ*.

REFERENCES

[1] "Bell System Technical Journal (1922–1983)" [Online]. Available: https://archive.org/details/bstj-archives&tab=about http://ieeexplore.ieee.org/xpl/RecentIssue.jsp?punumber=6731005 (accessed July 31, 2015).

[2] "Foreword," *The Bell System Technical Journal*, vol. 1, no. 1, pp. 1–3, July 1922. Also available online: http://ieeexplore.ieee.org/xpl/tocresult.jsp?isnumber=6773291&punumber=6731005 (accessed July 31, 2015).

[3] "Table of contents," *The Bell System Technical Journal*, vol. 62, no. 10, December 1983. Also available online: http://ieeexplore.ieee.org/xpl/mostRecentIssue.jsp?punumber=6731005 (accessed July 31, 2015).

(The original version of the column appeared in "AP-S turnstile," *IEEE Antennas and Propagation Magazine*, vol. 52, no. 6, p. 138, December 2010.)

NOTES

1. Bell Labs, for a long time the research arm of AT&T, became part of Lucent in 1996. Lucent was acquired by Alcatel in 2006 and now (December 2015) it seems that Lucent-Alcatel is being acquired by Nokia.
2. To learn more about Bell Labs and its role in American innovation, see Jon Gertner's book *The Idea Factory: Bell Labs and the Great Age of American Innovation*, Penguin, 2013.

1.9   AS SINGULAR AS A DELTA FUNCTION?

"Even in college, as I studied electrical engineering, I never worried very deeply about such things. What exactly is an electric or magnetic field? A dangerously shallow knowledge of quantum electrodynamics has clouded the issue, and the more I read the less I understand" [1].

R. Lucky

I am sure that many readers of Lucky's column share his trepidations (as I do) when it comes to making sense of quantum electrodynamics. Even P. A. M. Dirac (1902–1984), who shared the Nobel Prize for physics with Schrödinger in 1933, might have commiserated. Dirac, who received his first undergraduate degree in electrical engineering, felt that the quantum world could not be described in words or represented as images. To draw its picture would be *"like a blind man sensing a snowflake. One touch and it's gone"* [2]. He remarked on this challenge even in his Nobel Banquet Speech [3] of December 10, 1933: *"But the physicist is at a disadvantage in this respect on account of the very specialized nature of his work, which cannot be made intelligible without an intensive preliminary course of study."*

Interestingly, in the same Nobel Banquet Speech, given during the Great Depression, Dirac expressed his belief that *"[t]here is in my opinion a great similarity between the problems provided by the mysterious behavior of the atom and those provided by the present economic paradoxes confronting the world. In both cases one is given a great many facts, which are expressible with numbers, and one has to find the underlying principles. The methods of theoretical physics should be applicable to all those branches of thought, in which the essential features are expressible with numbers."* Clearly, in addition to anticipating antimatter particles, for example, a positron, Dirac was predicting the rise of the so-called "quants," the quantitative analysts (some of them trained as theoretical physicists) whose esoteric economic models might have precipitated (rather than solved, as Dirac had been hoping for) our recent "great recession." To quote another Nobel laureate (economics) Paul Krugman [4]: *"As I see it, the economics profession went astray because economists, as a group, mistook beauty, clad in impressive-looking mathematics, for truth."*

Let me be clear: quantum electrodynamics did not (I repeat not) create our recent economic problems and, furthermore, it is perfectly safe to delve deeper into the subject. So where does an electrical engineer start? A case can be made to go back to Dirac's own papers and his seminal book on the theory of electrons. After all, as Freeman Dyson, who took Dirac's course at Cambridge University, testified [5]: *"The great papers of the other quantum pioneers were more ragged, less perfectly formed than Dirac's. [Dirac's discoveries] were like exquisitely carved marble statues falling out of the sky, one after another. He seemed to be able to conjure laws of nature from pure thought."* Dyson's glowing recommendation notwithstanding, a more accessible source might well be *The Strangest Man*, a new biography of Dirac by G. Farmelo [6]. In addition to exploring the "hidden life" of Dirac, Farmelo places Dirac's work within the larger context of the developments in the twentieth century theoretical physics. Finally, even a cursory glance through the book will quickly reveal why the person who gave us the Dirac delta function might have been a true singularity himself.

REFERENCES

[1] R. Lucky, "The First Book of Electronics," *IEEE Spectrum* [Online]. Available: http://www.spectrum.ieee.org/geek-life/hands-on/the-first-book-of-electronics (accessed December 22, 2015).

[2] *The Times* (London) review of G. Farmelo's new biography of Dirac entitled *The Strangest Man* is available online at (subscription required): http://entertainment. timesonline.co.uk/tol/arts_and_entertainment/books/non-fiction/article5468057.ece.

[3] Dirac's Nobel Banquet Speech is available online at: http://nobelprize.org/nobel_ prizes/physics/laureates/1933/dirac-speech.html (accessed December 22, 2015).

[4] P. Krugman, "How Did Economists Get It So Wrong?" *The New York Times Magazine* [Online]. Available: http://www.nytimes.com/2009/09/06/magazine/06Economic-t.html?_r=1 (accessed December 22, 2015 2015).

[5] *The New York Times* review of G. Farmelo's new biography of Dirac entitled *The Strangest Man* is available at: http://www.nytimes.com/2009/09/13/books/review/Gilder-t.html (accessed December 22, 2015).

[6] G. Farmelo, *The Strangest Man*, Basic Books, 2009.

(The original version of the column appeared in "AP-S turnstile," *IEEE Antennas and Propagation Magazine*, vol. 51, no. 5, pp. 141–142, October 2009.)

NOTE

1. Many online resources discuss the Dirac delta function. See for example: http://mathworld.wolfram.com/DeltaFunction.html (accessed December 22, 2015)

1.10 PUBLISH OR PERISH?

While doing research on the Nobel laureate P. A. M. Dirac (1902–1984) for Section 1.9, I came across some interesting bits in the history of science. Let me start with a quiz:

What do Newton, Dirac, and Hawking have in common?

(a) All of them received a Nobel Prize for their contributions to physics.

(b) All of them have the same initial for their first names.

(c) All of them made use of Maxwell's equations in their work.

(d) None of the above.

The correct answer, as you must have guessed, is (d). So what is the common thread linking these justly celebrated physicists? They all occupied (at different times, of course) the Lucasian Chair of Mathematics at Cambridge University in England. Many readers may know that Maxwell (1831–1879) also taught at Cambridge University but he was *not* associated with the Lucasian Chair. Rather, in 1871, he became the first Professor of Experimental Physics and directed the newly created Cavendish Laboratory [1].

According to Reference [2], the Lucasian Chair of Mathematics was deeded in December 1663 at Cambridge University, England, as a result of a gift from Henry Lucas, Member of the Parliament (MP) for the university. The first holder of the Lucasian Chair was Isaac Barrow (1630–1677), who made fundamental theoretical contributions to geometrical optics. "… Barrow was required to lecture once a week during the term and to submit annually at least ten of these lectures to the vice-chancellor for deposit in the university library for public use." A strict publication requirement, if ever there was any! He was followed by Sir Isaac Newton as the second holder of the chair and then some 300 years later by Stephen Hawking, who occupied the chair for thirty years (1979–2009), retiring from the chair on 1 October, 2009 at the age of 67 (as required by the university rules). The next holder (2009–2013) of the chair was Michael Green, a co-founder of string theory [3].

Many holders of this prestigious academic chair have made important contributions to the field of electromagnetics. Here are some names one would readily recognize from graduate textbooks in the field:

- George Airy (1801–1892): Airy function
- George Stokes (1819–1903): Stokes' theorem
- Joseph Larmor (1857–1942): Larmor precession

As one would expect, holders of the Lucasian Chair have been prolific contributors to the scientific literature. For example, Airy's published papers have been counted at 377 in addition to another 141 official reports and addresses [2]. But to every rule, there must be an exception. Thus we come to the interesting case of Joshua King (1798–1857), who held the Lucasian Chair just before George Stokes. To quote from Joshua King's obituary [4]:

"His great mathematical power, however, did not lead him in the path of original investigation: with the exception of a short paper, containing 'A new demonstration of the Parallelogram of Forces,' read before the Cambridge Philosophical Society April 14, 1823, and published in Vol II of the Society's Transactions, we are not aware that he has left behind him any contribution to mathematical science."

Joshua King might have published little but at least he still enjoyed an enormous reputation at Cambridge throughout his tenure as the Lucasian Chair. Dirac was not so lucky. As noted in Reference [5], his last years as a Lucasian Chair were a bit of a trial:

"In post-war Cambridge, although still the Lucasian Professor, he was an irrelevance. They even took away his departmental parking space. Sick of such slights, Manci [his wife] persuaded him to accept an Eminent Professorship at Florida State University, where he became a revered curiosity."

REFERENCES

[1] For information on Maxwell, a good starting point is: http://www.clerkmaxwell foundation.org/index.html (accessed December 22, 2015).

[2] For information on the Lucasian chair, a valuable resource is: http://www-history.mcs.st-and.ac.uk/Societies/Lucasian.html (accessed December 22, 2015).

[3] The news of Michael Green's appointment to the Lucasian Chair is at: http://www. guardian.co.uk/science/2009/oct/20/stephen-hawking-michael-green-cambridge (accessed December 22, 2015).

[4] Joshua King's obituary is archived at: http://archive.is/D7x1u (accessed December 22, 2015).

[5] *The Times* (London) review of G. Farmelo's biography of Dirac entitled *The Strangest Man* is available online at (subscription required): http://entertainment. timesonline.co.uk/tol/arts_and_entertainment/books/non-fiction/article5468057.ece.

(The original version of the column appeared in "AP-S turnstile," *IEEE Antennas and Propagation Magazine*, vol. 51, no. 6, pp. 158–159, December 2009.)

DID YOU KNOW?

A FUN QUIZ (I)

Crème de la Crème

As a clutch of new science books [1] readily demonstrates, there is a lot of uncertainty not only about the current status of physics but also (Heisenberg's principle notwithstanding) about how fast the field is progressing (some would even say regressing). To be sure, there have been exciting developments [1] such as the experimental discovery of the Higgs boson, which completes the so-called Standard Model of particle physics, and the impressive new satellite data about the cosmic microwave background (CMB), which leads us ever closer to a deep understanding of the conditions prevailing in the early (i.e., shortly after the Big Bang) universe. Yet, when it comes to the "Holy Grail," namely a grand theory that would unify the Standard Model with Einstein's theory of general relativity, physicists seem to be as much at a loss now as they were nearly a generation ago. In his new book [2] *Farewell to Reality: How Modern Physics Has Betrayed the Search for Scientific Truth*, Jim Baggott complains bitterly about theoretical developments such as string theory, calling them "fairytale physics," since they have not been accompanied by actual experimental evidence or even testable predictions.

As Niels Bohr once quipped, "prediction is very difficult, especially if it is about the future." Perhaps, we can look in the rear-view mirror of history and agree more easily who the greatest contributors to physics have been before the current malaise set in. That was the task that the editors of *The Guardian* recently set for themselves when they looked for a list of the 10 best physicists [3]. To add to the fun, I have presented the list compiled by *The Guardian* in the format of a mini game show (think Jeopardy!). The answers appear at the bottom of this section.

1. Who supposedly received his inspiration about gravity from a falling piece of fruit?

2. Which Danish physicist came up with the modern idea of an atom, with a nucleus surrounded by revolving electrons?

3. Which Italian physicist got into trouble with the Vatican for his scientific theories?

4. Who came up with the famous equation for the equivalence of mass and energy?

5. Who discovered the theory of electromagnetism?

From ER to E.T.: How Electromagnetic Technologies Are Changing Our Lives, First Edition. Rajeev Bansal.
© 2017 by The Institute of Electrical and Electronic Engineers, Inc. Published 2017 by John Wiley & Sons, Inc.

6. Who discovered electromagnetic induction?

7. Who was the first woman to win a Nobel Prize?

8. Whose "diagrams" are used to illustrate quantum electrodynamic interactions?

9. After whom was the element rutherfordium named?

10. Who predicted the existence of antimatter but turned down a knighthood because he did not want people using his first name?

> One of (only?) the chief virtues of offering such a list is to get a conversation going among the readers. *The Guardian* not only anticipated such a debate but actually encouraged it by asking readers to submit their own choices. Many obliged and sent in what they considered serious omissions: Archimedes, Tesla, Planck, Heisenberg, Schrodinger, Lord Kelvin, Boltzmann, Pauli, _____ (you may enter your choices here)!

ANSWERS

1. Isaac Newton

2. Niels Bohr

3. Galileo Galilei

4. Albert Einstein

5. James Clerk Maxwell

6. Michael Faraday

7. Marie Curie

8. Richard Feynman

9. Ernest Rutherford

10. Paul Dirac

REFERENCES

[1] "Beyond the Numbers", *The Economist* [Online]. Available: http://www.economist.com/news/books-and-arts/21578366-fundamental-physics-has-made-important-advances-where-does-it-go-here-beyond (accessed April 21, 2016).

[2] J. Baggott, *Farewell to Reality: How Modern Physics Has Betrayed the Search for Scientific Truth*, Pegasus, 2013.

[3] "The 10 Best Physicists", *The Guardian* [Online]. Available: http://www.guardian.co.uk/culture/gallery/2013/may/12/the-10-best-physicists (accessed April 21, 2016).

(The original version of the quiz appeared in "AP-S turnstile," *IEEE Antennas and Propagation Magazine*, vol. 55, no. 3, pp. 176–177, June 2013.)

CHAPTER 2

THE EARTH AND BEYOND

"We live inside the ultimate high energy experiment."

—Neil Turok (1958–)

2.1 IN THE EYE OF THE BEHOLDER

"War es ein Gott, der diese Zeichen schrieb?"

(Was it a god who wrote these signs?)

From ER to E.T.: How Electromagnetic Technologies Are Changing Our Lives, First Edition. Rajeev Bansal.
© 2017 by The Institute of Electrical and Electronic Engineers, Inc. Published 2017 by John Wiley & Sons, Inc.

The quote [1] from Goethe's "Faust" above was physicist Boltzmann's (1844–1906) high tribute to the beauty of Maxwell's equations. Contemporary physicist Frank Wilczek, winner of the 2004 Nobel Prize, is equally entranced by Maxwell's elegant formulation of electromagnetic fields, where the situation [2] "takes on a life of its own, with the fields dancing as a pair, each inspiring the other." However, in Wilczek's case, like poet Browning's duchess [3], he likes whatever he looks on, and his looks go everywhere. In his new book [4] "A Beautiful Question: Finding Nature's Deep Design," Wilczek starts with the provocative question: "Is the world a work of art?" He continues [4]:

"Posed this way, our Question leads us to others. If it makes sense to consider the world as a work of art, is it a successful work of art? Is the physical world, considered as a work of art, beautiful? For knowledge of the physical world, we call on the work of scientists, but to do justice to our questions, we must also bring in the insights and contributions of sympathetic artists."

Wilczek recognizes immediately that the question has echoes of "spiritual cosmology." He elaborates on this historical connection by noting [4]:

"Galileo made the beauty of the physical world central to his own deep faith, as did Kepler, Newton and Maxwell. For all these searchers, finding beauty embodied in the physical world, reflecting God's glory, was the goal of their search. It inspired their work and sanctified their curiosity. And with their discoveries, their faith was rewarded."

However, as Wilczek is quick to point out [4], "while our Question finds support in spiritual cosmology, it can also stand on its own." He adds:

"And though its positive answer may inspire a spiritual interpretation, it does not require one. We will return to these thoughts toward the end of our meditation, by which point we will be much better prepared to appraise them. Between now and then, the world can speak for itself."

What comes "between now and then" is a chronological exploration of mankind's attempts to understand the beauty of the world with "symmetry" as the guiding principle. In an interview with the German magazine *Der Spiegel* [5], Wilczek explained:

SPIEGEL: Every artist has his or her own style. When you are investigating the laws of nature, do you feel that nature has its own style, too?

Wilczek: Absolutely. The world is a piece of art, produced according to a very peculiar style. What I find particularly striking is the outstanding role of symmetry.

SPIEGEL: Could you explain that?

Wilczek: Sure. The principle of symmetry as we use it in physics and mathematics can be described as "change without change." While this may sound mystical or bizarre, it actually means something quite simple. What, for instance, makes a circle such a symmetrical object? It's that you can rotate it around its center and it will remain a circle....This concept of symmetry as "change without change" can easily be generalized to laws of physics, or to the equations that express them.

Wilczek starts his exploration with the Greeks (Pythagoras) and continues the voyage through the current standard model of particle physics and the attempts to go beyond it though the principle of supersymmetry. If you choose to go on this bracing ride with Wilczek as your guide, keep in mind that the book can turn on a dime from its poetic language to a statement like "color gluons are the avatars of gauge symmetry 3.0." No wonder the book carries blurbs by both Lawrence Krauss and Deepak Chopra! [6]

REFERENCES

[1] My Collection of Quotes. The Cho Group. Department of Physics and Computer Science, Wake Forest University [Online]. Available: http://users.wfu.edu/choss/quotes.html (accessed October 17, 2015).

[2] L. Dartnell, "A Beautiful Question by Frank Wilczek: Review 'Worth the Effort'" *The Telegraph* [Online]. Available: http://www.telegraph.co.uk/culture/books/bookreviews/11773491/A-Beautiful-Question-by-Frank-Wilczek.html (accessed October 17, 2015).

[3] R. Browning, "My Last Duchess," *Poetry Foundation* [Online]. Available: http://www.poetryfoundation.org/poem/173024 (accessed October 17, 2015).

[4] F. Wilczek, A Beautiful Question: Finding Nature's Deep Design (excerpt) [Online]. Available: http://thepenguinpress.com/book/a-beautiful-question-finding-natures-deep-design/#excerpt (accessed October 17, 2015).

[5] J. Grolle, "Nobel Physicist Frank Wilczek: 'The World is a Piece of Art.'" *Der Spiegel* [Online]. Available: http://www.spiegel.de/international/physicist-frank-wilczek-interview-about-beauty-in-physics-a-1048669.html (accessed October 17, 2015).

[6] P. Ball, "A Physicist's Sense of Beauty," *PhysicsWorld* [Online]. Available (free registration required): http://physicsworld.com/cws/article/print/2015/oct/15/a-physicists-sense-of-beauty (accessed October 17, 2015).

(The original version of the column appeared in "Turnstile," *IEEE Antennas and Propagation Magazine*, vol. 58, no. 1, p. 96, February 2016.)

NOTES

1. The history of Maxwell's equations is presented by J. Rautio in the December 2014 issue of the IEEE Spectrum, available online at: http://spectrum.ieee.org/telecom/wireless/the-long-road-to-maxwells-equations (December 22, 2015).

2. Textbook resources:
   (i) W. H. Hayt and J. A. Buck, *Engineering Electromagnetics*, 8th ed., McGraw-Hill, New York, 2012. Maxwell's equations are introduced in Chapter 9.
   (ii) F. T. Ulaby and U. Ravaioli, *Fundamentals of Applied Electromagnetics*, 7th ed., Prentice Hall, Upper Saddle River, NJ, 2015. Maxwell's equations are presented in Chapter 6.

2.2   ROSES ARE RED, VIOLETS ARE BLUE...

One of the pleasures of teaching electromagnetics is to be able to demonstrate to students how this fundamental engineering science is not only responsible for designing hi-tech stuff like stealth aircraft, but also handy for explaining numerous natural phenomena. Take Rayleigh scattering, for example. Students in a course on radar learn that in the Rayleigh region, where the size of a spherical target is much smaller than the operating wavelength, the radar cross-section falls off as λ^{-4} and, consequently, [1] "rain and clouds are essentially invisible to radars which operate at relatively long wavelengths (low frequencies)." Of course, if it is a beautiful day outside, the same Rayleigh scattering may be invoked to explain why the sky is blue. The traditional explanation, which can be found in freshmen physics texts, is paraphrased by *The New York Times* as follows [2]: "Short-wavelength blue light bounces off more readily than long-wavelength red light, and that suffuses the sky with blue." Class dismissed!

As **Glenn Smith**, who is a Regents Professor Emeritus at Georgia Tech, noted in an article [3] in the *American Journal of Physics*, the above simple explanation has one fatal flaw: "From these considerations only [i.e., Rayleigh scattering], we could equally well say that the sky is violet." Or as the *Times* [2] article puts it, "why is the sky not purple?" Is there more going on here than meets the eye? The answer is yes, indeed; in fact, the color we perceive depends not only on what is going on in the atmosphere in terms of the scattering of light by the air molecules but also on what is going on inside the eye (color vision).

Smith shows that while the central notion that "color is not a property of light itself, but is a sensation produced by the human visual system" can be traced all the way back to Newton's *Opticks*, a detailed understanding of the process is relatively recent. Within the human retina, specialized photoreceptors called cones (some 5 million of them) are responsible for processing color information. There are three types of cones; each contains a pigment that makes that group sensitive to a specific range of wavelengths. Among them, they cover (in an overlapping fashion) the visible part of the spectrum. The signals from *all* three types of cones are integrated (with respect to the wavelength) and summed to produce the perceived color response. One point to note here is the similarity to an "inverse problem"; the output (the perceived color) can result from multiple combinations of inputs (incident light). For example, a suitable combination of red and green lights will make the observer see yellow. In the case of the sky, it turns out that the combination of the scattered light wavelengths excites the cones in a way that is interpreted by the eye/brain as a mixture of pure blue and pure white, which the observer calls light (sky) blue.

In addition to quantitative calculations to support his explanation, Smith's paper also includes a careful description, complete with a list of parts and their suppliers, of a demonstration suitable for classroom use. Moreover, the appendix includes a couple of problems for students who really want to master the material and the extensive list of references covers the work done over the past 300 years. When I asked Smith how he got going on this exhaustive investigation of color vision, he explained [4]:

> "When I wrote my book [5] on classical electromagnetic radiation, I prepared a chapter on dipole radiation. As an example, I included dipole scattering from molecules and a description of how this caused the blue color and polarization of skylight. At the time, I realized that the eye's response was also an important factor, but I only mentioned this in a footnote. Later, I decided to really look into the problem, and I obtained a number of books and papers on color vision. I got hooked! Over a period of a few years, I refined my lectures on this material and developed experiments for classroom demonstration. Finally, I decided to write the paper for the American Journal of Physics (AJP), thinking that others might like to use this material in their classes."

So to go back to the nursery rhyme, are violets really blue? As Smith has convincingly demonstrated, the answer really lies in the eye of the beholder.

REFERENCES

[1] M. Skolnik, *Introduction to Radar Systems*, 2nd ed., McGraw-Hill, 1980.

[2] K. Chang, "Deep Purple = Moody Blues," *The New York Times*, July 19 2005.

[3] G. Smith, "Human color vision and the unsaturated blue color of the daytime sky," *American Journal of Physics*, vol. 73, no. 7, pp. 590–597, July 2005.

[4] G. Smith, private communication, 2005.

[5] G. Smith, *An Introduction to Classical Electromagnetic Radiation*, Cambridge University Press, 1997.

(The original version of the column appeared in "AP-S turnstile," *IEEE Antennas and Propagation Magazine*, vol. 47, no. 4, pp. 128–129, February 2005.)

NOTE

1. Textbook resources:
   (i) W. H. Hayt and J. A. Buck, *Engineering Electromagnetics*, 8th ed., McGraw-Hill, New York, 2012. Dipole radiation is discussed in Chapter 14.

(ii) F. T. Ulaby and U. Ravaioli, *Fundamentals of Applied Electromagnetics*, 7th ed., Prentice Hall, Upper Saddle River, NJ, 2015. Dipole radiation is discussed in Chapter 9. The basic operation of a radar is presented in Chapter 10.

2.3 2003: AN EARTH ODYSSEY?

Navigators all at sea

Don't eat onions for their tea

Not that they're at all emetic

They make the compass nonmagnetic

As I noted in Reference [1], "British naval helmsmen were once flogged if they were found to be in violation of the regulation that 'steersmen, and such as tend the Mariner's Card are forbidden to eat Onyons and Garlick, lest they make the index of the poles drunk.'" In the twenty-first century, when the compass fails, the source of the problem may not be onboard in the form of onion-eating steersmen, but, as viewers of the Hollywood thriller *The Core* may suspect, it may lie deep inside the Earth's magnetic core. Sending a probe to the center of the Earth to tackle the problem may look great on celluloid but it will surely invite only derision in scientific circles. Or will it?

Let's look at the premise first. According to an article by geophysicist J. Marvin Herndon in the *Proceedings of the National Academy of Sciences*, the geomagnetic field is doomed indeed because the nuclear reactor at the center of the earth is running down. Other geophysicists consider any obituary of the earth's magnetic field premature by at least a few billion years. Whew! [2].

As for the solution, we turn to CalTech geophysicist David Stevenson's "modest proposal" in *Nature* [3]. In the sixteenth century, as readers of Reference [1] may recall, Gilbert constructed a spherical lodestone called terrella ("little earth") for his work on geomagnetism. Using small compass needles, he explored the magnetic field of his terrella and used his findings to estimate the effect of the earth's magnetic field on the behavior of a magnetic compass. While geophysicists today have a much more sophisticated suite of tools to study geomagnetism, Stevenson feels that, in comparison with billions of dollars spent on unmanned space exploration, funding for studying the earth's interior has been minimal. Hence his modest proposal:

> *"Here I propose a scheme for a mission to the Earth's core, in which a small communication probe would be conveyed in a huge volume of liquid-iron alloy migrating down to the core along a crack that is propagating under the action of gravity. The grapefruit-sized probe would transmit its findings back to the surface using high-frequency seismic waves sensed by a ground-coupled wave detector. The probe should take about a week to reach the core, and the minimum mass of molten iron required would be 10^8–10^{10} kg — or roughly between an hour and a week of Earth's total iron-foundry production."*

Of course, the crack will have to be initiated. Stevenson suggests using the equivalent of a few megatons of TNT (presumably from a nuclear device) to create an underground explosion with the force of a magnitude-7 earthquake. University of Connecticut geophysicist Vernon Cormier noted [4] that Stevenson's proposal "allows the hole to keep open because it is not really an open hole or an open crack – it is a crack filled by a fluid that is denser than its surroundings."

Stevenson realized that his proposal might sound far-fetched, but he said [4]: "I will be happy even if people just think it's funny, but I'll be happier still if some people take it somewhat seriously."

I would like to conclude the column with a little thought experiment drawn from Robert Bank's entertaining book [5], *Slicing Pizzas, Racing Turtles, and Further Adventures in Applied Mathematics*. Stevenson's probe is expected to take about a week to travel 1800 miles to the core of the Earth. Banks ponders the hypothetical question of traveling under the influence of gravity down an evacuated shaft drilled through the earth. Simple calculations reveal that the one-way trip *all* the way across to the other side will take a mere 42 minutes and 10 seconds. And yes, you would be zipping along at Mach 23 as you pass through the center of the Earth. Now that's a speed that even Major Rebecca Childs (Hilary Swank) of *The Core* won't consider too shabby for her terra-ship.

REFERENCES

[1] R. Bansal, "AP-S turnstile: De Magnete," *IEEE Antennas and Propagation Magazine*, vol. 42, p. 110, October 2000.

[2] *U. S. News & World Report,* March 17, 2003.

[3] D. Stevenson, "Mission to Earth's core – a modest proposal," *Nature*, vol. 423, pp. 239–240, May 15, 2003.

[4] B. Cosgrove-Mather, "Journey to the Center of the Earth," *CBSNews* [Online]. Available: http://www.cbsnews.com/news/journey-to-the-center-of-the-earth-14-05-2003/ (accessed December 19, 2015).

[5] R. Banks, *Slicing Pizzas, Racing Turtles, and Further Adventures in Applied Mathematics*, Princeton University Press, 1999.

(The original version of the column appeared in "AP-S turnstile," *IEEE Antennas and Propagation Magazine*, vol. 45, no. 3, p. 134, June 2003.)

NOTE

1. Reference [1] is included in this book as Section 1.6.

2.4 WHICH CAME FIRST: BIG BANG OR BIG CRUNCH?

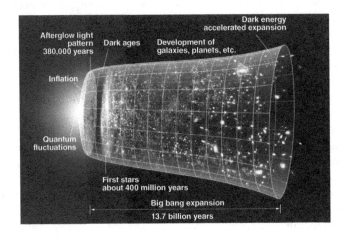

"We live inside the ultimate high energy experiment."

—Neil Turok
Cambridge University

On the first day of my graduate class in microwave engineering, I try to remind the students that microwave radio astronomy has been one of the driving forces behind the spectacular success of the "standard" account of the origin of our universe, popularly known as the inflationary Big Bang theory. How Bell Labs scientists Arno Penzias and Robert Wilson almost accidentally discovered in 1965 that the electromagnetic noise picked up by their horn antenna was not caused by a "white dielectric substance" (pigeon droppings) within the antenna, but represented the cosmic microwave background (CMB) makes for an exciting story even after all the intervening years. The CMB was found to be diffuse, coming essentially uniformly from all directions, and had an equivalent temperature of around 2.7 K. Since it originated *only* 300,000 years after the Big Bang, the genesis moment around 14 billion years ago when space and time began, it provides crucial experimental data about the early history of our universe.

Since the pioneering work of Penzias and Wilson, several teams around the world have been making increasingly sophisticated microwave measurements to unravel the finer details of the CMB. For example, the Cosmic Background Imager (CBI), located at an altitude of 16,700 feet in the dry Chilean desert, consists of 13 radio antennas operating at frequencies from 26 to 36 GHz. A similar interferometer, Degree Angular Scale Interferometer (DASI), operates from the NSF Amundsen–Scott South Pole station. These instruments have detected minute variations (as small as 10 millionths of a degree) in the CMB, corresponding to "the first tentative seeds of matter and energy that would later evolve into clusters of hundreds of galaxies." The measured temperature fluctuations in the CMB also support the concept of an "inflationary period," a modification of the Big Bang theory. According to the inflationary Big Bang theory, the nascent universe underwent an extreme and rapid expansion (inflation) in the first 10^{-32} seconds, which made the universe homogeneous (on a large scale) and flat (parallel lines do not intersect) and "which created the fluctuations that seeded the formation of galaxies and large-scale structure."

While the cosmological observations of the last decades have provided solid support for the inflationary Big Bang theory, they have also required some *ad hoc* fixes to the theoretical model. According to Paul Steinhardt of Princeton University, "it appears that, billions of years after the big bang, following the formation of galaxies, the Universe was overtaken by some form of dark energy that is causing the expansion rate to accelerate." While most cosmologists have been happy simply to graft "dark energy" on to the prevailing inflationary Big Bang model, others such as Steinhardt of Princeton and Turok of Cambridge University have chosen to go back to the drawing board and conceive of a new paradigm—*the cyclic universe*. As Steinhardt explains it, "in this picture, space and time exist for ever. ... The Universe undergoes an endless sequence of cycles in which it contracts in a big crunch and re-emerges in an expanding big bang, with trillions of years of evolution in between." The centerpiece of the new paradigm is the transition from the big crunch to big bang. Employing the framework of a variation of string theory known as M theory (used in particle physics), Steinhardt et al. conceive of the universe as two branes (three-dimensional surfaces) oscillating along a (hidden) extra dimension, with the collision corresponding to the transition from the big crunch (the collapse of the extra dimension) to big bang (emergence of a new universe). (An animation of the cyclic universe is available on the websites of Steinhardt and Turok.)

Stephen Hawking once implied that the end of theoretical physics might be in sight. I tend to agree with Turok's recent observation that "the prospects of linking cosmology to fundamental theory are bright and that even if the End isn't in sight, the Beginning might be!"

SOURCES

"Microwave Imager Probes Universe 'First Light' to Answer Cosmological Questions," NSF PR 02-41, May 23, 2002.

S. Begley, "Latest Observations Steal the Thunder From Big Bang Theory," *The Wall Street Journal*, April 26, 2002.

R. Cowen, "When branes collide," *Science News*, vol. 160, no. 12, p. 184, September 22, 2001.

J. Glanz, "Listen Closely: From Tiny Hum Came Big Bang," *The New York Times*, April 30, 2001.

P. Steinhardt, The Endless Universe: A Brief Introduction to the Cyclic Universe [Online]. Available: http://www.physics.princeton.edu/~steinh/philo/philo2.htm (accessed December 19, 2015).

N. Turok, The Origins of Our Universe. Posted at Professor Turok's website at Cambridge University (UK) [Online]. Available: http://www.damtp.cam.ac.uk/user/ngt1000/ (accessed December 19, 2015).

(The original version of the column appeared in "Microwave surfing," *IEEE Microwave Magazine*, vol. 2, no. 2, pp. 32–34, December 2002.)

NOTE

1. Nobel laureate Steven Weinberg's short book *The First Three Minutes* (updated edition: Basic Books, 1993) remains a fine entry point for those interested in learning more about cosmology (including the Big Bang theory and the CMB).

2.5 WHISTLING IN THE DARK?

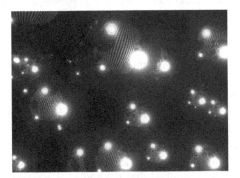

Bell Labs scientists Arno Penzias and Robert Wilson received the 1978 Nobel Prize in physics for their "accidental" discovery in 1965 that "the electromagnetic noise picked up by their horn antenna was not caused by a "white dielectric substance" (pigeon droppings) within the antenna, but represented the CMB. The CMB was found to be diffuse, coming essentially uniformly from all directions, and had an equivalent temperature of around 2.7 K. Since it originated *only* 300,000 years after the big bang (the genesis moment around 14 billion years ago when space and time began), it provides crucial experimental data about the early history of our universe" [1].

Nearly half a century after the pioneering work of Penzias and Wilson, radio measurements of the sky at microwave frequencies continue to make waves. In order to explain the radiometric observations at frequencies ranging from 3 to 90 GHz, obtained from NASA's ARCADE 2 mission, Professor Fornengo of the University of Turin in Italy and his colleagues have invoked the role of extragalactic dark matter [2–4].

As described on NASA's website, "The Absolute Radiometer for Cosmology, Astrophysics, and Diffuse Emission (ARCADE) ... consists of a set of 7 precision radiometers, cooled to nearly absolute zero, and carried to an altitude of over 35 km (21 miles) by a scientific research balloon." ARCADE had a successful flight on July 22, 2006 and the payload, which included a full complement of radiometers, was carried to an altitude of 37 km (120,000 feet) for approximately 4 hours of observation. The mission "detected a radio background much brighter than expected. The detected background is 5 to 10 times too bright to be explained as the combined emission from distant radio galaxies" [3].

What could be the source of this excess radio noise? In a provocative paper [4], Professor Fornengo's European team ties it to dark matter (DM). As for the dark matter [5], "we are much more certain what dark matter is not than we are what it is. First, it is dark, meaning that it is not in the form of stars and planets that we see. ...Second, it is not in the form of dark clouds of normal matter, matter made up of particles called baryons. ...Third, dark matter is not antimatter, because we do not see the unique gamma rays that are produced when antimatter annihilates with matter. Finally, we can rule out large galaxy-sized black holes on the basis of how many gravitational lenses we see." Elusive as it may be, dark matter has been essential to fitting a theoretical model of the composition of the universe (~70% dark energy, ~25% dark matter, ~5% normal matter) to the combined set of cosmological observations [5].

In the thesis advanced by Professor Fornengo's group, "the ARCADE excess is the existence of lots and lots of very faint sources, rather like the makeup of 'haloes' of dark matter outside galaxies. When dark-matter particles—known as weakly interacting massive particles or WIMPs—collide and annihilate one another, they are thought to generate electrons and positrons, which subsequently generate radio waves via synchrotron emission as they travel through magnetic fields" [2].

Not all astrophysicists feel that an exotic scenario involving DM is warranted to explain the ARCADE data. Professor Fornengo readily agrees that as additional radiometric data, for example, from the Square Kilometer Array (SKA), becomes available, it would become clearer whether his group's DM hypothesis is correct or whether a more banal explanation would suffice. Stay tuned.

REFERENCES

[1] R. Bansal, "Which came first: big bang or big crunch?" *IEEE Microwave Magazine*, vol. 3, no. 4, pp. 32–34, December 2002. DOI: 10.1109/MMW.2002.1145673

[2] J. Cartwright, "Radio-Wave Excess Could Point to Dark Matter" *PhysicsWorld* [Online]. Available: http://physicsworld.com/cws/article/news/48018 (accessed December 19, 2015).

[3] ARCADE mission website [Online]. Available: http://arcade.gsfc.nasa.gov/ (accessed December 19, 2015).

[4] N. Fornengo, R. Lineros, M. Regis, and M. Taoso, "A Dark Matter Interpretation for the ARCADE Excess?" [Online preprint]. Available: http://arxiv.org/abs/1108.0569 (accessed December 19, 2015).

[5] NASA's astrophysics website [Online]. Available: http://science.nasa.gov/astrophysics/focus-areas/what-is-dark-energy/ (December 19, 2015).

(The original version of the column appeared in "AP-S turnstile," *IEEE Antennas and Propagation Magazine*, vol. 53, no. 6, pp. 162–163, June 2011.)

NOTES

1. Reference [1] is included in this book as Section 2.4.
2. To learn more about The Square Kilometer Array (SKA) see Section 3.1.
3. Nobel laureate Steven Weinberg's short book *The First Three Minutes* (updated edition: Basic Books, 1993) remains a fine entry point for those interested in learning more about cosmology (including the Big Bang theory and the CMB).

2.6 GOING BEYOND A SELFIE

Ultrahigh-energy cosmic rays, with energies [1] of 10^{20} eV or more when they strike the atmosphere, are both mysterious in origin and fiendishly difficult to study because of their rare occurrence. They pack [2] "the energy of a rapidly delivered cricket ball into a single, blisteringly fast atomic nucleus" and "each square kilometer of Earth is hit, on average, by about one a century. The Pierre Auger Observatory, a facility in Argentina which surveys 3000 square kilometers of Earth's atmosphere, picked up around 15 a year between 2005 and 2008." No wonder unusual proposals are being floated to try to detect these ultrahigh-energy bursts of cosmic rays. One [2] by British astrophysicist Justin Bray, which I mentioned in Reference [3], would use the moon as a huge cosmic ray detector and then rely upon the giant radio telescope the SKA to listen for the brief radio-frequency pulse produced when the cosmic ray burst enters the lunar regolith.

If Bray's proposal sounds too much like a moon shot, here is one that can be implemented using a device that almost everyone carries in his pocket: a smartphone. The idea [1] here is to use the CMOS chip in the smartphone's camera to detect the secondary particles produced when cosmic rays collide with air molecules in the Earth's atmosphere. Here is how it is supposed to work:

> "All you need to get started is to download the app from the Internet [4], plug your phone in and then place it face-up on an opaque surface. With the camera obscured in this way, the CMOS pixels are more or less shielded from visible photons but remain exposed to particles from cosmic-ray showers, such as higher-energy photons and muons, which can pass through walls, tables and plastic phone cases to generate a measurable voltage when they ionize atoms of silicon. The resulting pattern of bright pixels is stored within the phone as a stream of data, which is then uploaded to a central server to be analysed whenever a wireless Internet connection becomes available" [1].

You may be wondering that, if ultrahigh-energy bursts of cosmic rays are such rare events that even vast facilities like the Pierre Auger Observatory register only a few per year, what chance a lowly smartphone would have of capturing such an event. This is where it gets interesting. Apple alone has sold over 700 million iPhones since 2007 [5] and the iPhone is not even the most common smartphone around the world. What if we could get some of the millions of smartphone users to join together in a collective search for cosmic rays? Daniel Whiteson (University of California, Irvine) and Michael Mulhearn (University of California, Davis) ran some numbers and this is what they came up with:

> "[T]o be confident of intercepting almost every shower produced by cosmic rays with an energy of at least 10^{20} eV, some 1000 activated phones would be needed in each square kilometre of coverage – the idea being that at least five phones should register a hit within five seconds of one another in order to distinguish signals from (uncorrelated) noise. The researchers also calculated how big an area they would need to cover in order to match Pierre Auger's 'exposure' – the product of its observing area, field of view and duration of data taking – and the answer they came up with was 825 km^2. They therefore concluded that the total number of phones needed would be around 825,000" [1].

"Crowdsourcing" in the service of science has been tried before. For example, since 1999, when the free screensaver program SETI@home [5] was released, many volunteers worldwide have donated the idle time on their computers to sift through the radio signals collected by the Search for Extraterrestrial Intelligence (SETI) telescopes. Skeptics caution though that, if a popular program such as SETI currently has only around 120,000 active users [1], the number of smartphone users participating in the search for cosmic rays may not hit the needed number (say 825,000). Others question many of the technical assumptions made by Whiteson and Mulhearn in coming up with their smartphone network model. All I can say is that it sounds like fun to join the global band of cosmic ray hunters.

REFERENCES

[1] E. Cartlidge, "Dialling Up the Cosmos" *PhysicsWorld* [Online]. Available: http://physicsworld.com/cws/article/print/2015/jan/15/dialling-up-the-cosmos (accessed March 9, 2015).

[2] "Moonbeams," *The Economist* [Online]. Available: http://www.economist.com/news/science-and-technology/21621705-intriguing-proposal-study-cosmic-rays-looking-earths (accessed March 9, 2015).

[3] R. Bansal, "AP-S turnstile: the annual quiz," *IEEE Antennas and Propagation Magazine*, vol. 57, no. 1, February 2015.

[4] CRAYFIS app website [Online]. Available: http://crayfis.io/about (accessed March 9, 2015).

[5] J. Dove, "Cook: iPhone 6 Sales Grew at 'Double the Industry Rate.'" *TNW* [Online]. Available: http://thenextweb.com/apple/2015/03/09/cook-iphone-6-sales-grew-at-double-the-industry-rate/ (accessed March 9, 2015).

(The original version of the column appeared in "Turnstile," *IEEE Antennas and Propagation Magazine*, vol. 57, no. 2, pp. 77–78, April 2015.)

NOTES

1. Reference [3] is included in the book as *A Fun Quiz (ii)*.
2. SETI@home and the Square Kilometer Array (SKA) are discussed in Section 3.1.

DID YOU KNOW?

A FUN QUIZ (II)

1. If you really and truly want to escape the modern world of cell phones and Wi-Fi, consider moving to _____.

   (a) Caliente, NV

   (b) Green Bank, WV

   (c) Whitehall, MT

   (d) None of the above

2. 2015 marked the 150th anniversary of the publication by James Clerk Maxwell of his eponymous equations for electromagnetism. Maxwell's original formulation of his electromagnetic theory contained _____ equations.

   (a) 4

   (b) 8

   (c) 20

   (d) None of the above

3. The Nobel Prize in Physics 2014 was awarded in the area of "applied physics" for the invention of the blue light-emitting diode (LED). Which physics discipline attracts the most Nobel prizes?

   (a) Atomic, molecular, and optical physics

   (b) Condensed matter physics

   (c) Nuclear and particle physics

   (d) None of the above

4. British astrophysicist Justin Bray planned to study rare bursts of ultrahigh-energy cosmic rays by using the _____ as a giant cosmic ray detector.

   (a) Moon

   (b) Earth

   (c) Sun

   (d) None of the above

From ER to E.T.: How Electromagnetic Technologies Are Changing Our Lives, First Edition. Rajeev Bansal.
© 2017 by The Institute of Electrical and Electronic Engineers, Inc. Published 2017 by John Wiley & Sons, Inc.

5. Sony filed a patent for a smart _____ that will be able to process data, communicate wirelessly, and help monitor physiological parameters such as blood pressure.
   (a) Sweater
   (b) Tie
   (c) Wig
   (d) None of the above

6. Drexel University researchers have developed a _____ antenna sensor for measuring uterine contractions in pregnant women.
   (a) Knitted
   (b) Printed
   (c) Nano
   (d) None of the above

7. Jonathan Cheseaux, a Swiss computer science student, has developed a system that lets a drone find natural disaster survivors through their _____.
   (a) Smart watches
   (b) Cell phones
   (c) Google glasses.
   (d) None of the above

8. The 2013 Isaac Newton Medal was awarded to the British physicist John Pendry, who is known to our professional community for his work on _____.
   (a) Evolutionary computational techniques
   (b) Magnetic monopoles
   (c) Metamaterials
   (d) None of the above

9. The historic first technical meeting of the American Institute of Electrical Engineers (a predecessor to the IEEE) in _____ in 1884 was attended by Bell, Edison, and Tesla.
   (a) Boston
   (b) New York
   (c) Philadelphia
   (d) None of the above

10. What washes up on tiny beaches?
   (a) Nanoparticles
   (b) Microwaves
   (c) Mini-algae
   (d) None of the above

ANSWERS

1. **(b)** Green Bank, WV
   Source: M. Gaynor, "The Town Without Wi-Fi," *Washingtonian* [Online]. (*Note*: I would like to thank Ric Tell, Chair of the IEEE Committee on Man and Radiation (COMAR), for bringing this item to my attention. Green Bank is the site of the Robert C. Byrd Radio Telescope, operated by the National Radio Astronomy Observatory, and is within the National Radio Quiet Zone, where radio transmissions are strongly restricted by law.) Available: http://www.washingtonian.com/articles/people/the-town-without-wi-fi/?src=longreads (accessed January 9, 2015).

2. **(c)** 20
   Source: J. Rautio, "The Long Road to Maxwell's Equations. *IEEE Spectrum*" [Online]. (*Note*: It was Oliver Heaviside who reduced the original 20 equations to the modern 4 in 1885.) Available: http://spectrum.ieee.org/telecom/wireless/the-long-road-to-maxwells-equations (accessed January 9, 2015).

3. **(c)** Nuclear and particle physics
   Source: H. Johnston, "What Type of Physics Should You Do If You Want to Bag a Nobel Prize?" *PhysicsWorld* [Online]. (*Note*: 35 Nobel Prizes have been shared by 68 laureates in the area of nuclear and particle physics.) Available: http://blog.physicsworld.com/2014/10/02/what-type-of-physics-should-you-do-if-you-want-to-bag-a-nobel-prize/ (accessed January 9, 2015).

4. **(a)** Moon
   Source: "Moonbeams," *The Economist* [Online]. (*Note*: Bray plans to use the giant radio telescope the Square Kilometer Array (SKA) to listen for the brief radio-frequency pulse produced when the cosmic ray burst enters the lunar regolith.) Available: http://www.economist.com/news/science-and-technology/21621705-intriguing-proposal-study-cosmic-rays-looking-earths (accessed January 9, 2015).

5. **(c)** Wig
   Source: "Wearable Tech," *The IET* [Online]. Available: http://www.theiet.org/resources/inspec/support/subject-guides/wearable-tech.cfm (accessed January 9, 2015).

6. **(a)** Knitted
   Source: "Wearable Technology" [Online]. Available: http://drexelnanophotonics.com/wearable-technology/ (accessed January 9, 2015).

7. **(b)** Cell phones
   Source: "A Drone That Finds Survivors Through Their Phones" [Online]. Available: http://actu.epfl.ch/news/a-drone-that-finds-survivors-through-their-phones/ (accessed January 9, 2015).

8. **(c)** Metamaterials

   Source: H. Johnston, "John Pendry Wins 2013 Isaac Newton Medal." *PhysicsWorld* [Online]. Available: http://physicsworld.com/cws/article/news/2013/jul/01/john-pendry-wins-2013-isaac-newton-medal (accessed January 9, 2015).

9. **(c)** Philadelphia

   Source: A. Davis, "IEEE Milestone Recognizes the AIEE's First Technical Meeting," *The Institute* [Online]. Available: http://theinstitute.ieee.org/technology-focus/technology-history/ieee-milestone-recognizes-the-aiees-first-technical-meeting (accessed January 9, 2015).

10. **(b)** Microwaves

    Source: I owe this "pun" to one of my daughters. ☺

(The original version of the quiz appeared in the "Turnstile," *IEEE Antennas and Propagation Magazine*, vol. 57, no. 1, pp. 66–70, February 2015.)

CHAPTER 3

SEARCH FOR EXTRATERRESTRIAL INTELLIGENCE (SETI)

"A time will come when men will stretch out their eyes—they should see planets like our Earth."

—Christopher Wren (1632–1723)

3.1 LITTLE GREEN MEN: A PHANTOM MENACE?

In movie theaters across the country, there was a (p)reincarnation of sorts in 1999; I am talking about the release of *Star Wars: Episode I—The Phantom Menace*, a prequel to the highly successful *Star Wars* series. Of course, Hollywood was not the only place interested in the continuing saga of our intergalactic neighbors. Around

From ER to E.T.: How Electromagnetic Technologies Are Changing Our Lives, First Edition. Rajeev Bansal.
© 2017 by The Institute of Electrical and Electronic Engineers, Inc. Published 2017 by John Wiley & Sons, Inc.

the same time, a few hundred miles to the north, at UC Berkeley, the Department of Astronomy appointed a brand new chair in the Search for Extraterrestrial Intelligence (SETI). William "Jack" Welch (now a professor emeritus) was the first holder of the Watson and Marilyn Alberts Chair in SETI as well as the then vice president of the Mountain View based SETI Institute.

One of Welch's projects, dubbed the One Hectare Telescope at the time (and now known as the Allen Telescope Array), was supposed to be a phased array of hundreds of radio telescopes, representing a total effective aperture of 10,000 square meters, or 1 hectare (about 2.5 acres). The project cost? A modest $25 million for such a huge system. According to Welch, "by using a large number of satellite TV antennas and inexpensive receivers we will build ourselves, we can get a very sensitive antenna that is much cheaper than the cost of building one big reflecting dish and one large receiver."

Welch's wife, Jill Tarter, a scientist at the SETI Institute and the inspiration for Jodie Foster's character in the film *Contact*, cast an even wider net. She was all excited about the potential development of the Square Kilometer Array (SKA), comprising antennas scattered over an area of 1 million square kilometers, with an effective aperture diameter of about 1000 km. The astronomers dreaming about the SKA talk mainly about the chance to observe the "epoch of first light," an early stage in the evolution of the universe. However, to Tarter, the half-billion dollar SKA was also a great opportunity to watch interstellar TV. She figured that the SKA will be powerful enough to detect the leakage into space from a standard terrestrial TV transmitter at a range of half a dozen light years. Therefore, any neighboring civilization with comparable TV transmitters would leak enough signal into space for detection by the SKA, even if it were not intentionally broadcasting for our benefit. (I wonder what we would think if they turned out to be watching reruns of the original *Star Wars*.)

Can't wait till these grandiose projects are fully operational? No problem. Thanks to the SETI@home project, which was launched on May 17, 1999, anyone with a home computer and an internet connection can immediately participate in the worldwide search for SETI. You log on to the site SETI@home, download a piece of software, and leave the rest to your computer. When the computer is "idle," the program goes to work analyzing the data collected by the actual SETI program. The data comes from the Arecibo radio telescope in Puerto Rico, which is searching (part time) for possible alien signals. After the computer is finished, it sends the results back to SETI@home scientists at UC Berkeley and grabs another chunk of data. Welcome to the world of the grand distributed experiment!

SOURCES

Based on reports from the BBC, the *Houston Chronicle*, the *Hartford Courant*, the *Economist*, and the Spring '99 issue of the UC Berkeley *EECS/ERL News*.

(The original version of the column appeared in "AP-S turnstile," *IEEE Antennas and Propagation Magazine*, vol. 41, no. 3, p. 90, June 1999.)

NOTES

1. For more information about the Allen Telescope Array, consult the website: http://www.seti.org/ata (accessed April 27, 2016).
2. To learn more about the Square Kilometer Array project, see: https://www.skatelescope.org/ (accessed April 27, 2016).
3. The technical challenges involved in watching reruns of old TV shows across interstellar distances are discussed at: http://contactincontext.org/lucy.htm (accessed April 27, 2016).
4. More details about the SETI@home project are available at the project website at: http://setiathome.ssl.berkeley.edu/ (accessed April 27, 2016).
5. To learn more about the Arecibo radio telescope, which supplies the data used by SETI@home, consult the website of the Arecibo Observatory at http://www.naic.edu/general/ (accessed April 27, 2016).

3.2 WAITING FOR GODOT?

"A time will come when men will stretch out their eyes—they should see planets like our Earth" [1].

—Sir Christopher Wren (1632–1723)

Over the last three centuries, there have been many advances in optical telescopes. To find planets that may support life like our own planet, astronomers have to search for planets that orbit stars at the "Goldilocks" distance; not so close that they will be unbearably hot "Jupiters" and not so far away that they will be frozen "Plutos." Locating objects within the bright glare of a host star is no easy task. But, as reported in a paper [1, 2] in *Nature* by a Jet Propulsion Laboratory (JPL) group, scientists have made tremendous progress in the field. The JPL group used "wavefront correction" techniques applied to coronagraphs to observe optically a planet orbiting its host 33 light years away with a relatively small (1.5 m) earth-based telescope.

However, when it comes to searching for extraterrestrial intelligence (SETI), the tool of choice for the last 50 years has been a radio telescope rather than an optical one. Almost 1000 star systems have been scrutinized for "intelligent" radio signals using increasingly sophisticated phased-array antennas. The Allen Telescope Array (funded largely by Paul Allen, co-founder of Microsoft) in California currently has 42 dish antennas, each 6 m in diameter. It is eventually supposed to grow to 350 dishes and would be able to observe one million star systems within a decade [3].

Despite five decades of progressively sophisticated radio listening, all we have heard are the sounds of silence. Like the characters in the absurdist play "Waiting for Godot," are we waiting in vain? Now some scientists are beginning to question the whole premise of radio-based SETI [4]. Discussing his new book on the subject of SETI [5], British physicist Paul Davies notes [6]:

> "A fundamental flaw lies at the core of most existing SETI strategies. Carl Sagan popularized the appealing idea that an altruistic alien community might be obligingly beaming radio messages at us, perhaps carefully crafted to give mankind a welcome technological and sociological fillip. But that scenario will no longer wash. Even SETI optimists concede that a radio-savvy civilization within a few hundred light years is extremely unlikely (and systematic searches have spotted nothing). Suppose there is an alien community 1,000 light years away. That is still in our galactic neighborhood – the Milky Way is some 100,000 light years across. The aliens belonging to this putative community cannot know of our existence – they cannot know that Earth has radio technology and the means to detect their signals. The reason concerns the finite speed of light. At 1,000 light years away, the aliens see Earth today as it was 1,000 years ago. Because nothing can go faster than light (it is a basic law of physics), there is no way they can know about the industrial revolution and terrestrial radio telescopes. So why would they have started beaming messages to us 1,000 years ago, when their view of Earth at that time would have been the year A.D. 10? They might detect signs of agriculture and large scale building (such as the pyramids), and they may of course surmise that some millennium soon humans would develop radio technology. But it would make no sense for them to start transmitting powerful and expensive radio messages at us until they know we are on the air. When will that be? In about 900 years time, when our first feeble radio transmissions, leaking into space at the speed of light, finally reach them."

Not everyone is anxiously awaiting close encounters with extraterrestrial beings. Physicist Stephen Hawking, in a series for the Discovery channel, said [7] that while it was "perfectly rational" to assume the existence of intelligent life elsewhere, humans should do everything possible to avoid contact. He cautions: "If aliens visit us, the outcome would be much as when Columbus landed in America, which didn't turn out well for the Native Americans.... We only have to look at ourselves to see how intelligent life might develop into something we wouldn't want to meet." So maybe it is just as well that Godot hasn't shown up.

REFERENCES

[1] "Extrasolar Planets: A trick of the Light," *The Economist* [Online]. Available: http://www.economist.com/sciencetechnology/displaystory.cfm?story_id=15905845 (accessed December 23, 2015).

[2] E. Serabyn, D. Mawet, and R. Burrus, "An image of an exoplanet separated by two diffraction beamwidths from a star," *Nature*, vol. 464, pp. 1018–1020, 2010.

[3] "Signs of Life," *The Economist*, pp. 89–90, April 17, 2010.

[4] "Seeking Extraterrestrial Intelligence: A Deathly Hush," *The Economist* [Online]. Available: http://www.economist.com/books/displaystory.cfm?story_id=15864923 (accessed December 23, 2015).

[5] P. Davies, *The Eerie Silence: Renewing Our Search for Alien Intelligence*, Houghton Mifflin Harcourt, 2010.

[6] Paul Davies discussed his book [5] at: http://www.amazon.com/Eerie-Silence-Renewing-Search-Intelligence/dp/0547133243 (accessed December 23, 2015).

[7] "Hawking Warns over Alien Beings," *BBC* [Online]. Available: http://news.bbc.co.uk/2/hi/uk_news/8642558.stm (accessed December 23, 2015).

(The original version of the column appeared in "AP-S turnstile," *IEEE Antennas and Propagation Magazine*, vol. 52, no. 3, pp. 124–125, June 2010.)

NOTE

1. Section 3.1 provides additional links to radio-based SETI.

3.3 IS THERE ANYBODY THERE?

"'Is there anybody there?' said the Traveller,

Knocking on the moonlit door;

And his horse in the silence champed the grasses

Of the forest's ferny floor" [1].

—Walter de la Mare (1873–1956)

Like the bemused traveler in Walter de la Mare's celebrated poem, we have been wondering for a long time if anyone else is out there in the universe. In a recent testimony before a House Committee on Science, Seth Shostak, senior astronomer at the SETI Institute [2], offered the following optimistic assessment:

"It's unproven whether there is any life beyond Earth. *I think that situation is going to change within everyone's lifetime in this room*" [3].

As I reported in Reference [4], despite five decades of progressively sophisticated radio listening (thousands of star systems have been scrutinized for "intelligent" radio signals using increasingly sophisticated phased-array antennas), all we have heard are the sounds of silence. So what accounts for Shostek's prophetic language at the Congressional hearing? Let us look at the "evidence" [3] he offered at his testimony.

- NASA's Kepler telescope has revealed an abundance of planets in the galaxy suggesting that there may be tens of billions of potentially habitable ("earth cousins") planets in the Milky Way alone. Shostak noted, "If this is the only planet on which not only life, but intelligent life, has arisen, that would be very unusual."

- While determining the success rate of a radio listening program such as SETI is difficult, the best estimates suggest that one needs to examine a few *million* star systems in order to have a reasonable chance of success. Shostak expects that, as radio telescope technology advances, that number will be achievable. It requires vast computing resources to analyze the data received by the radio telescopes to look for "intelligent" signals. Since 1999, when the free screensaver program SETI@home [5] was released, millions of volunteers worldwide have donated the idle time on their computers to sift through the radio signals collected by the SETI telescopes.

- Another approach to the search for life beyond Earth involves looking for evidence for *microbial* life (present or past) on other planets or their moons. Robotic missions to Mars are one current example of this approach. As Shostak observed, there are "at least half a dozen other [potentially habitable] worlds" within our solar system.

- Another approach involves examining the atmospheres of planets orbiting other stars for evidence of gases such as oxygen and methane that correlate with biological activity. When a planet passes between Earth and *its* sun, a thick enough atmosphere may be potentially observable.

Shostak testified that the last two approaches (the search for microbial life and the analysis of planetary atmospheres) could yield persuasive data within the next two decades.

While I remain enthusiastic about a multi-pronged search for life beyond Earth, others, including physicist Stephen Hawking, have cautioned [4] that although it was "perfectly rational" to assume the existence of intelligent life elsewhere, humans should do everything possible to avoid contact. Such skeptics (cynics?) would draw comfort from the following lines that come toward the end of Walter de la Mare's poem [1] that I quoted from above:

"For he suddenly smote on the door, even

Louder, and lifted his head:–

'Tell them I came and no one answered,

That I kept my word,' he said."

REFERENCES

[1] Walter de la Mare, "The Listeners" [Online]. Available: http://www.poetryfoundation.org/poem/177007 (accessed May 31, 2014).

[2] The SETI Institute website [Online]. Available: http://www.seti.org/ (accessed May 31, 2014).

[3] "Will We Meet Aliens in the Near Future?," *The Christian Science Monitor* [Online]. Available: http://www.csmonitor.com/layout/set/print/Science/2014/0528/Will-we-meet-aliens-in-the-near-future-video (accessed May 31, 2014).

[4] R. Bansal, "AP-S turnstile: Waiting for Godot," *IEEE Antennas and Propagation Magazine*, vol. 52, no. 3, pp. 124–125, June 2010.

[5] SETI@home [Online]. Available: http://setiathome.ssl.berkeley.edu/ (accessed May 31, 2014).

(The original version of the column appeared in "AP-S turnstile," *IEEE Antennas and Propagation Magazine*, vol. 56, no. 3, pp. 186–187, June 2014.)

NOTE

1. Reference [4] is included in this book as Section 3.2.

3.4 SCIENCE OR SCIENCE FICTION?

In his 1937 science-fiction novel Star Maker, the British philosopher and writer Olaf Stapledon [1] imagined an advanced civilization that built a spherical shell around its home star to capture all the emitted photons in order to meet its vast energy needs. Such a sphere would render the star invisible to the rest of the universe *except*, as the Princeton physicist Freeman Dyson noted [2] in 1960, the sphere (known now as the

"Dyson sphere" [3]) itself would reradiate the heat generated by the civilization into outer space in the mid-infrared range. Furthermore, in a brief article [4] in *Science*, Dyson proposed that *"a search for sources of infrared radiation should accompany the recently initiated search for interstellar radio communications"* [in looking for clues to the existence of extraterrestrial civilizations]. In response to readers' comments that a solid shell surrounding a star might be mechanically impossible, Dyson clarified [4]: *"The form of 'biosphere' which I envisaged consists of a loose collection or swarm of objects traveling on independent orbits around the star. The size and shape of the individual objects would be chosen to suit the inhabitants. I did not indulge in speculations concerning the constructional details of the biosphere, since the expected emission of infrared radiation is independent of such details."*

Let us fast forward to the twenty-first century. Funded by a grant from the New Frontiers program of the John Templeton Foundation, astronomer Jason Wright and his colleagues at Pennsylvania State University have been pursuing the project G-HAT: Glimpsing Heat from Alien Technologies search for extraterrestrial intelligence [5]. They have been making use of NASA's Wide-Field Infrared Explorer (WISE) space telescope for their search. Wright cheerfully explains [6]: *"WISE was launched by NASA for pure, natural astrophysics; it just happened to be perfect Dyson sphere finder."* A search of 100,000 galaxies for the "tell-tale" infrared signatures of advanced civilizations has not found any evidence to date but the team has found some puzzling new sources of mid-infrared radiation that merit further study. As Wright observes [7] more careful future observations will, "push our sensitivity to alien technology down to much lower levels, and to better distinguish heat resulting from natural astronomical sources from heat produced by advanced technologies. This pilot study is just the beginning."

Distinguishing artifacts from the desired signals has been a longstanding challenge for astronomers. A recent unrelated development [8] involving an Australian astronomers group is a case in point. The team working at the Parkes Observatory has found that the source of at least one kind of perytons (millisecond-duration transients of terrestrial origin whose frequency-swept emission mimics fast radio bursts of extragalactic origin) turned out to be microwave ovens being used at the observatory. The group noted [8] in a recently uploaded manuscript: *"Subsequent tests revealed that a peryton can be generated at 1.4 GHz when a microwave oven door is opened prematurely and the telescope is at an appropriate relative angle. Radio emission escaping from microwave ovens during the magnetron shut-down phase neatly explain all of the observed properties of the peryton signals."*

REFERENCES

[1] "Olaf Stapledon, 1886–1950," *eBooks@Adelaide* [Online]. Available: https://ebooks. adelaide.edu.au/s/stapledon/olaf/ (accessed April 6, 2015).

[2] "Infra Digging," *The Economist* [Online]. Available: http://www.economist.com/news/ science-and-technology/21648607-search-extraterrestrials-goes-intergalactic-infra-digging (accessed April 6, 2015).

[3] D. Byrd, "What is a Dyson Sphere?" *EarthSky* [Online]. Available: http://earthsky.org/space/what-is-a-dyson-sphere (accessed April 6, 2015).

[4] F. Dyson, "Search for Artificial Stellar Sources of Infrared Radiation (abstract)," *Science* [Online]. Available: http://www.islandone.org/LEOBiblio/SETI1.HTM (accessed April 6, 2015).

[5] J. Wright, "The G Search for Kardashev Civilizations," *AstroWright* [Online]. Available: http://sites.psu.edu/astrowright/the-g-hat-search-for-kardashev-civilizations/ (accessed April 6, 2015).

[6] T. Lewis, "Incredible Technology: How to Search for Advanced Alien Civilizations," *Livescience* [Online]. Available: http://www.livescience.com/42540-how-to-search-for-alien-civilizations.html (accessed April 6, 2015).

[7] T. Commissariat, "Trail Runs Cold on Alien Hotspots, for Now," *Physicsworld* [Online]. Available: http://physicsworld.com/cws/article/news/2015/apr/28/trail-runs-cold-on-alien-hotspots-for-now (accessed April 6, 2015).

[8] R. Yirka, "Mystery of Peryton Reception at Australian Observatory Solved: It's From Microwave Ovens," *Phys.Org* [Online]. Available: http://phys.org/news/2015-04-mystery-peryton-reception-australian-observatory.html (accessed April 6, 2015).

(The original version of the column appeared in "Turnstile," *IEEE Antennas and Propagation Magazine*, vol. 57, no. 3, p. 15, June 2015.)

NOTE

1. In fall 2015, there was considerable excitement in the SETI community when it was reported that the star KIC 8462852, located 1400 light years away, periodically dimmed its normal brightness. Were there orbiting solar energy collectors (so-called Dyson swarms) that were blocking the light? The SETI Institute turned its Allen Telescope array in the star's direction to search for radio transmissions. Alas, no such signal was found. As Seth Shostak, Senior Astronomer at the SETI Institute noted, "while it's nice to hope for megastructures, don't bet the family farm on it." http://www.seti.org/seti-institute/news/are-there-signals-coming-deep-space (accessed April 28, 2016).

  The Breakthrough Listen initiative, funded primarily by Russian billionaire Yuri Milner, is planning to make additional observations of the star in 2017. Stay tuned. http://spectrum.ieee.org/transportation/mass-transit/100-million-breakthrough-listen-initiative-starts-searching-for-et

DID YOU KNOW?

A FUN QUIZ (III)

1. Florida International University researchers are taking principles from the traditional Japanese art of origami to create powerful, yet compact _____.
   - **(a)** Unmanned aerial vehicles (UAVs)
   - **(b)** Antennas
   - **(c)** Submarine sails
   - **(d)** None of the above

2. The well-known superheterodyne receiver, which uses frequency mixing or heterodyning to convert a received signal to a more convenient fixed frequency *lower* than that of the original radio signal, was invented in 1917 by _____.
   - **(a)** R. Fessenden
   - **(b)** E. Armstrong
   - **(c)** L. Levy
   - **(d)** None of the above

3. Researchers at the University of Texas, Arlington have developed a _____, a device about 1/10 the size of a single grain of rice, that (the researchers hope) may one day help power your smart phone.
   - **(a)** Fuel cell
   - **(b)** Nuclear cell
   - **(c)** Micro-windmill
   - **(d)** None of the above

4. In 1997, as part of a research study on diamagnetism, Andre Geim and his colleagues used strong magnetic fields to levitate a live frog. Their work won the 2000 Ig Nobel Prize in physics (administered by the *Annals of Improbable Research*). Geim went on to win the 2010 Nobel Prize in physics for his research on _____.
   - **(a)** Graphene
   - **(b)** Superconductivity
   - **(c)** Fractional Hall effect
   - **(d)** None of the above

---

5. _____ had the *wireless* feature of his implanted defibrillator disabled so that nobody could attempt to assassinate him by hacking into the device.

   (a) Former US President Clinton

   (b) Former US Secretary of Defense Rumsfeld

   (c) Former US Vice President Cheney

   (d) None of the above

6. For future fifth generation (5G) cellular networks, researchers at New York University and Samsung are hoping to tap into the _____ frequency band.

   (a) Millimeter wave

   (b) Terahertz

   (c) Very high frequency (VHF)

   (d) None of the above

7. Physicists in Spain and Germany have designed and built a "magnetic hose" from ferromagnetic and superconducting materials to _____.

   (a) Create a noiseless vacuum cleaner

   (b) Transmit magnetic fields

   (c) Extract magnetic ore from the ground

   (d) None of the above

8. The average person looks at his/her phone about _____ times a day, thus providing a great opportunity to get the right information to people at the right time in the fast-developing field of "mHealth."

   (a) 15

   (b) 60

   (c) 150

   (d) None of the above

9. Atacama Large Millimeter/sub-millimeter Array (ALMA), constructed at a cost of $1.3 billion in _____, consists of 66 dishes and is considered the world's most powerful radio telescope.

   (a) Hawaii

   (b) Australia

   (c) Chile

   (d) None of the above

10. The experimental multi-static primary surveillance radar (MSPSR) in the UK relies upon _____ signals reflected by aircraft to track them.

    (a) Cellular communication

    (b) Broadcast television

    (c) Broadcast radio

    (d) None of the above

ANSWERS

1. (b) Antennas
Source: "FIU Engineers Use Origami to Design Antennas," *NBC* [Online]. Available: http://www.nbcmiami.com/news/FIU-Engineers-Use-Origami-to-Design-Antennas-242370551.html (accessed January 5, 2016).

2. (c) L. Levy
Source: A. Douglas, "Who Invented the Superheterodyne?" [Online]. (*Note*: I would like to thank F. Broyde and E. Clavelier for bringing the controversy surrounding this invention, commonly attributed to Armstrong, to my attention.) Available: http://antiqueradios.com/superhet/ (accessed January 5, 2016).

3. (c) Micro-windmill
Source: W. Pentland, "Micro-Windmills May One Day Power Your Smart Phone," *Forbes* [Online]. Available: http://www.forbes.com/sites/williampentland/2014/01/10/micro-windmills-may-one-day-power-your-smart-phone/ (accessed January 5, 2016).

4. (a) Graphene
Source: R. Wesson, "The Wow Factor," *Prism* [Online]. Available: http://www.asee-prism.org/discovery/ (accessed January 5, 2016).

5. (c) Former US Vice President Cheney
Source: N. Gass, "Dick Cheney Feared Assassination by Heart-Device Hack," *Politico* [Online]. Available: http://www.politico.com/story/2013/10/dick-cheney-feared-assassination-by-heart-device-hack-98550.html (accessed January 5, 2016).

6. (a) Millimeter wave
Source: D. Talbot, "What 5G Will Be: Crazy-Fast Wireless Tested in New York City," *Technology Review* [Online]. Available: http://www.technologyreview.com/news/514931/what-5g-will-be-crazy-fast-wireless-tested-in-new-york-city/ (accessed January 5, 2016).

7. (b) Transmit magnetic fields
Source: E. Cartlidge, "Introducing the Magnetic Hose," *PhysicsWorld* [Online]. Available: http://physicsworld.com/cws/article/news/2013/may/01/introducing-the-magnetic-hose (accessed January 5, 2016).

8. (c) 150
Source: B. Cleland, "mHealth Moving Along" [Online]. Available: http://www.ipi.org/policy_blog/detail/mhealth-moving-along (January 5, 2016).

9. **(c)** Chile

   Source: "The Great Test Tube in the Sky," *The Economist* [Online]. Available: http://www.economist.com/news/science-and-technology/21573533-space-one-big-chemistry-set-great-test-tube-sky (accessed January 5, 2016).

10. **(b)** Broadcast television

    Source: "A Programme Worth Watching," *The Economist* [Online]. Available: http://www.economist.com/news/science-and-technology/21573527-how-air-traffic-control-can-use-television-signals-plot-aircraft-programme (accessed January 5, 2016).

(The original version of the quiz appeared in "AP-S turnstile," *IEEE Antennas and Propagation Magazine*, vol. 56, no. 1, pp. 200–203, February 2014.)

CHAPTER 4

PROFESSIONALISM: ETHICS AND LAW

4.1 DID MAXWELL PULL A FAST ONE?

Some time ago I received a radiometer as a gift (was it father's day? birthday?) from my children, who are always trying to add—shall we say—visual interest to my office space. A radiometer, also known as a light-mill and more precisely as a Crookes Radiometer, in case you have not come across one before, is billed as an educational toy or a novelty ornament and sold by science-oriented outlets as well

From ER to E.T.: How Electromagnetic Technologies Are Changing Our Lives, First Edition. Rajeev Bansal.
© 2017 by The Institute of Electrical and Electronic Engineers, Inc. Published 2017 by John Wiley & Sons, Inc.

as, in less expensive versions, by online retailers such as Amazon. It consists of an airtight glass bulb mounted on a pedestal. Inside the glass bulb, a set of four upright metallic vanes is mounted on thin horizontal arms projecting from a vertical spindle. The vanes are silvery on one side and dark on the other. When the radiometer is placed in light (mine sits on the windowsill, where it catches plenty of sunlight), the vanes begin to turn, with the silvery side of the vanes leading the way. The brighter the light (natural or artificial), the faster is the rotation. No batteries required.

When I opened the gift at the dinner table, my children did a little demo for me and then asked me if I knew how it worked. I had not seen a radiometer before (really) but I thought (and said as much) that the radiometer was an elegant demonstration of the radiation pressure exerted by light. The photons striking the reflective silvery side exerted more pressure than those being absorbed by the absorptive dark side. Case closed. Next?

After dinner, I decided to "confirm" my clear-cut explanation by (what else) googling "radiometer." Boy, was I ever wrong. As I learned from Reference [1], this was the explanation offered by the Victorian chemist Sir William Crookes, who developed the radiometer in 1873 while investigating infrared radiation and the element thallium. Crookes submitted a paper using differential radiation pressure to describe the operation of the radiometer. The paper was reviewed by Maxwell, who accepted Crookes's explanation, taking delight apparently in the experimental vindication of radiation pressure predicted by his electromagnetic theory. However, within a few years, it was noted that if radiation pressure were responsible for making the vanes turn, the vanes would turn in the direction *opposite* to what is actually observed. The brighter reflective side would have the greater radiation pressure and, therefore, the vane would be pushed toward the darker side. (Much later in 1901, it was found that if the glass bulb had a perfect vacuum, the radiometer would not work at all, indicating that the gases inside the glass bulb played some role in the operation of the Crookes radiometer.)

In 1879, Osborne Reynolds (remember the "Reynolds number" from fluid dynamics?) offered the correct *qualitative* explanation for the radiometer's rotation based on his theory of "thermal transpiration" in a manuscript submitted to the Royal Society. In the presence of a thermal gradient, gas molecules exert tangential forces along the edges of a vane, which results in the rotation observed. Maxwell was again a reviewer for the manuscript; he bought Reynolds's explanation but recommended changes to the analysis. Then, shortly before his death in November 1879, Maxwell's own detailed mathematical analysis appeared in a paper "On stresses in rarefied gases arising from inequalities of temperature" in the *Philosophical Transactions of the Royal Society*. Maxwell's paper acknowledged Reynolds's (unpublished) work but criticized his (unpublished) mathematical treatment. Reynolds's own paper was not published till 1881. He wanted the Royal Society to publish his protest against Maxwell's questionable conduct, "but after Maxwell's death this was deemed inappropriate."

By the way, a radiometer (Nichols radiometer) based on the principle of radiation pressure was developed by Nichols and Hull in 1901 [2]. The original apparatus is at the Smithsonian. If I ever receive a replica as a gift, I promise to tell you more about it.

REFERENCES

[1] P. Gibbs, "How Does a Light-Mill Work?" [Online]. Available: http://math.ucr.edu/home/baez/physics/General/LightMill/light-mill.html (accessed January 5, 2016).

[2] D. Lee, "A Celebration of the Legacy of Physics at Dartmouth," *Dartmouth Undergraduate Journal of Science*. Dartmouth College, 2008 [Online]. Available: http://dujs.dartmouth.edu/spring-2008-10th-anniversary-edition/what-else-has-happened-a-celebration-of-the-legacy-of-physics-at-dartmouth (accessed January 5, 2016).

(The original version of the column appeared in "AP-S turnstile," *IEEE Antennas and Propagation Magazine*, vol. 52, no. 5, p. 186, October 2010.)

NOTES

1. For a mathematical derivation of radiation pressure, see for example: http://farside.ph.utexas.edu/teaching/em/lectures/node90.html (accessed April 27, 2016).

2. To learn more about the ethics of peer review, see for example: http://research-ethics.net/topics/peer-review/#discussion (accessed April 27, 2016).

4.2 CELL PHONES AND CANCER: ANATOMY OF A LEGAL OPINION

Background

Claiming that his use of an analog cell phone manufactured by Motorola caused his brain cancer, Dr. Christopher Newman and his wife filed an $800 million lawsuit in 2000 against Motorola, two wireless industry associations, and several wireless service providers. He was represented by a prominent Baltimore law firm, which has successfully sued asbestos companies in the past. Although Dr. Newman's case was not the first instance of a lawsuit alleging a link between cell phone usage and brain cancer, it attracted a lot of attention in the wireless industry because it advanced the farthest. Early in 2002, the US District Court for the District of Maryland directed both parties to begin the "discovery" phase focused on the issues of both *general* and *specific* causation: that is, (1) can the use of wireless handheld telephones cause brain cancer?, and (2) did the use of the Motorola phone cause Dr. Newman's cancer? A preliminary evidentiary hearing was held in February and March 2002 and post-hearing correspondence was received by the judge through September 2002. Later that month, Judge Catherine Blake dismissed the lawsuit. Her detailed memorandum, discussed below, is useful for the insight it provides on the *legal* admissibility of scientific evidence.

Federal Rules of Evidence

Citing *Daubert* v. *Merrell Dow* (1993), Judge Blake noted that the trial court must perform a two-pronged analysis in order to satisfy its gate-keeping function. According to her, "the first question is whether scientific evidence is *valid and reliable*. The second question is whether it will help the trier of fact, which is generally a question of *relevance* or 'fit': assuming the evidence is reliable, does it apply to the facts in the individual case under consideration." In the Daubert case, the Court had identified several factors that may bear on the determination of admissibility of scientific evidence. These include (in Judge Blake's words):

1. whether a theory or techniques can be or has been tested;
2. whether it has been subjected to peer review and publication;
3. whether a technique has a high known or potential rate of error and whether there are standards controlling its operation;
4. whether the theory or technique enjoys general acceptance within a relevant scientific community.

Verdict

After reviewing the testimony from both sides, Judge Blake concluded that the causation opinions (cell phone usage causes brain cancer) offered by Dr. Newman's experts did *not* pass the Daubert test. She felt that "the reasoning, theories, and methodology [supporting the plaintiffs] have not gained general acceptance in the scientific community, as demonstrated by the numerous national and international scientific and governmental published reports finding no sufficient proof that use of handheld cellular phones causes human brain cancer, and by the array of established, experienced, and highly-credentialed experts called to testify by the defense. The only published peer-reviewed epidemiological study finding such causation has serious flaws, and reliable epidemiology is essential before any link between animal studies and human cancer causation can be made. Neither Dr. Hardell's [Swedish epidemiological] work nor Dr. Lai's animal studies, heavily relied on by the plaintiffs' experts, have been replicated or otherwise validated by other scientists. Further Dr. Lai's published studies [performed at 2.45 GHz] lack relevance, or 'fit,' when applied to RFR or cell phone frequency [around 850 MHz for analog phones]."

Dr. Newman's attorney, John Angelos, conceded: "We didn't pass the standard….It's pretty much a complete victory [for the defendants]."

Postlude

The Supreme Court's opinion on the issue of the admissibility of evidence probably bears repetition here:

> "Conclusions and methodology are not entirely distinct from one another…A court may conclude that there is simply too great an analytical gap between the data and the opinion proffered."

SOURCES

Memorandum (2002) by Judge Blake in the case of *Newman et al.* v. *Motorola et al.*
The Wall Street Journal, October 1, 2002.
WIRED NEWS, September 30, 2002.
Boston Herald, September 30, 2002.

(The original version of the column appeared in "Microwave surfing," *IEEE Microwave Magazine*, vol. 4, no. 1, pp. 28–30, March 2003.)

NOTES

1. The putative health hazards of cell phone radiation are discussed also in Section 5.1.
2. The current position of the US Food and Drug Administration (FDA) on the safety of cell phones is posted at: http://www.fda.gov/Radiation-EmittingProducts/Radiation EmittingProductsandProcedures/HomeBusinessandEntertainment/CellPhones/ucm116282 .htm (accessed April 27, 2016).
3. For more information on the *Daubert* v. *Merrell Dow* case, see for example: http:// www.casebriefs.com/blog/law/torts/torts-keyed-to-prosser/causation-in-fact/daubert-v-merrell-dow-pharmaceuticals-inc-4/ (accessed April 27, 2016).

4.3 HAPPY 200TH ANNIVERSARY!

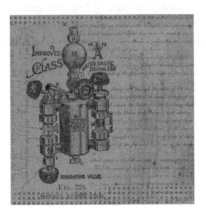

In 2002, as the US Patent and Trademark Office (USPTO) celebrated its 200th anniversary [1], I compiled the following vignettes from the colorful history of patents.

The USPTO has its roots in the US Constitution, which empowered Congress "to promote the Progress of Science and useful Arts, by securing for limited Times to Authors and Inventors the exclusive Right to their respective Writings and Discoveries." Under the original patent law [2] of 1790, patent applications were to be examined by none other than the secretary of state, the secretary of war, and the attorney general. Thus, Thomas Jefferson (then secretary of state) was one of the distinguished examiners for US patent #1, awarded on July 31, 1790 for an improvement "in the making [of] Pot Ash and Pearl Ash by a new Apparatus and Process." Pot Ash (potash) was America's first industrial chemical [3]. Only two other inventions were deemed worthy of patents that year. In contrast, in 2002, the USPTO had 3300 examiners, who issued around 3500 patents and 2000 trademarks each *week* [1]. As for quality, modern patents range from US patent #6,000,000 for "hot synchronization" technology for handhelds awarded to 3Com to US patent #5,965,809 for a method to determine the bra size by direct measurement of the user's chest [2].

* * *

The earliest radio patent (US patent #129,971) was awarded to Dr. Mahlon Loomis on July 20, 1872 for an improvement to (wireless) telegraphy. Dr. Loomis had demonstrated "potential differences on a galvanometer between two kites during a lightning storm, 14 miles apart in Loudoun County, Virginia in October 1866." In 1900, Marconi received his famous patent #7777, which incorporated:

- Use of aerial and ground
- Inductive coupling to the aerial and ground circuits

- Use of tuning coils to obtain the desired wavelength
- Employed the electrical energy of the earth as a battery [4]

* * *

In 1905, while Einstein was a Patent Office clerk in Zurich, he produced (in his spare time) three seminal papers for a single volume of *Annalen der Physik* [2]. Spare time must be hard to come by for modern USPTO examiners, who often have to search 500,000 documents and review as many as 160,000 pages of information submitted with a patent application within the space of 20 hours [1].

* * *

Compared to Thomas Edison's 1093 patents, Bill Gates holds only 1 patent. It is estimated [1] that a laptop may incorporate components covered by as many as 5000 patents. That number does not include patent #2,524,035 (expired) for "Semiconductor Amplifier; Three-Electrode Circuit Element Utilizing Semiconductive Materials" [2] (the original Bell Labs transistor).

* * *

In the mid-nineteenth century, Abraham Lincoln, the only US president to hold a patent (#6469 for a "Device for Buoying Vessels over Shoals") felt that "the patent system added the fuel of interest to the fire of genius." Fifty years later, the US Supreme Court expressed a more cynical view when it noted: "It creates a class of speculative schemers who make it their business to watch the advancing wave of improvement, and gather its foam in the form of patented monopolies, which enable them to lay a heavy tax upon the industry of the country." Another 100 years later, as the USPTO churns out patents at a pace unimaginable to its founders, we are still debating whether every good idea deserves a 20-year government-sponsored monopoly. It is worth remembering that one of the most influential inventions of our time "the world-wide web" was *not* patented by its inventor Tim Berners-Lee, who has worked hard to keep the system open and non-proprietary [2].

* * *

REFERENCES

[1] The patenting process was the central theme of the *Forbes ASAP* (Summer 2002) issue.

[2] The James Gleick article "Patently Absurd" (*The New York Times Magazine*, March 12, 2000) is available online at www.around.com (accessed January 5, 2016).

[3] The information about the first patent granted by the USPTO is available on many websites including http://inventors.about.com/od/weirdmuseums/ig/Inventive-Thinking/First-Patent-Grante.htm (accessed January 5, 2016).

[4] The information about the early history of radio patents is available at http://smart90.com/nbstubblefield (accessed January 5, 2016).

(The original version of the column appeared in "AP-S turnstile," *IEEE Antennas and Propagation Magazine*, vol. 44, no. 4, p. 86, August 2002.)

4.4 EINSTEIN DOESN'T WORK HERE ANYMORE

I am told that agent Fox Mulder, the fictional FBI agent in the popular TV series *The X-Files*, who specializes in investigating "extraordinary" cases, has a poster on his office wall that reads: I WANT TO BELIEVE. Read the following "exposition" and see if you want to be a believer.

"All known radio transmissions use known models of time and space dimensions for sending the RF signal.

The present invention has discovered the apparent existence of a new dimension capable of acting as a medium for RE signals. Initial benefits of penetrating this new dimension include sending RF signals faster than the speed of light, extending the effective distance of RF transmitters at the same power radiated, penetrating known RF shielding devices, and accelerating plant growth exposed to the by-product energy of the RF transmissions.

The following describes, in simple terms, what the present invention actually does. The present invention takes a transmission of energy, and instead of sending it through normal time and space, it pokes a small hole into another dimension, thus, sending the energy through a place which allows transmission of energy to exceed the speed of light.

The following is a description of how the communications medium converter functions.

First, you need to create a hot surface that is more than 1000 degrees Fahrenheit. Next, it requires a strong magnetic field. Then, you need an accelerator, followed by an electromagnetic injection point. For communications or data communication, you need 2 devices. Each device is connected to a transmitter and receiver. This allows electromagnetic energy to enter a dimension and to travel at speeds faster than the speed of light.

The magnetic fields are focused onto the heat generating device. The electromagnetic injection point is the plane generated by the two opposing magnetic fields.

It has been observed by the inventor and witnesses that accelerated plant growth can occur using the present invention."

My guess is that the typical reader's response to the above description will be reminiscent of German-born mathematician Richard Courant's (1888–1972) words: "Zis is not nuts. Zis is supernuts." The US Patent Office, on the other hand, must have wanted to believe, because in February 2000, it issued Patent #6,025,810 to David L. Strom of Aurora, CO for a "hyper-light-speed antenna."

How could the Patent Office be persuaded by something that sounds so weird to issue a patent? One explanation was suggested by Attorney Lee Henderson in his June 2000 column of the *IEEE Antennas and Propagation Magazine*: "It seems better to allow a few such patents to slip through, in order to make sure that groundbreaking inventions are properly recognized. History is replete with examples of inventions that were initially dismissed because they challenged the existing theories."

A somewhat different perspective is offered by Philip Ross in a *Forbes* article "Patently Absurd." He quotes patent expert Gregory Aharonian: "In mechanical arts, a lot of examiners have been around for decades and know the technology like the backs of their hands. Chemistry has great databases for prior art." However, when it comes to other areas—electronics, software—the examiners could surely use someone like Einstein on their side (after all, Einstein got his start in a Swiss patent office). Moreover, the sheer volume of work being handled by the US Patent Office is staggering. According to *Forbes*, "last year [1999] 3,200 examiners awarded 161,000 patents, or 50 apiece…One software reviewer approved 200 last year. That is four new patents each week."

Of course, even Einstein might have been intrigued by the results of a recent experiment purporting to show that light, under the right circumstances, can travel faster than c. In the experiment, described in a paper submitted to *Nature* by Dr. L. Wang of NEC Research Institute in Princeton, NJ, a pulse of light traversing a chamber of specially prepared cesium gas "appears to be pushed to speeds of 300 times the normal speed of light. That is so fast that, under these peculiar circumstances, the *main* part of the pulse exits the chamber even before it enters" (*The New York Times*). However, when the experiment is analyzed in terms of group velocity and backward traveling modes, most physicists feel that the experiment does *not* represent the transmission of information at a speed greater than c, and, therefore, no fundamental principles are violated. Whew!

SOURCES

Forbes (May 29, 2000).
The New York Times (May 30, 2000).

The Courant quote is from a book review in the June 4, 2000 issue of the Sunday *NY Times*.

The Mulder poster story appears in *Voodoo Science* by Robert L. Park (Oxford University Press, August 2000).

https://patents.google.com/ (accessed April 27, 2016).

(The original version of the column appeared in "AP-S turnstile," *IEEE Antennas and Propagation Magazine*, vol. 42, no. 4, pp. 104–105, August 2000.)

NOTE

1. **Bernard E. Souw** of the US Patent and Trademark Office wrote to me in October 2000 and offered the following comments on the hyper-light-speed antenna (US Patent #6,025,810) discussed in the August 2000 column (reproduced above):

   "I was told by the Patent Examiner who allowed the Strom's patent, that, although the specification of the 'invention' describes a method for 'generating RF signals faster than light,' the claims themselves do not contain, or have been cleaned from, such wordings. Therefore, I was further told, the invention as claimed is legally patentable. As far as I can ascertain from the specification and the claims, the invention as claimed would work, i.e., generating RF signals, although the signal propagation will–of course–not travel faster than c. However, your concern is well accepted, and I have forwarded a copy of your article (with my comments) to the corresponding Art Unit."

(Excerpted from my column "AP-S turnstile," *IEEE Antennas and Propagation Magazine*, vol. 42, no. 6, pp. 118–119, December 2000.)

4.5 IT IS A BIRD. IT IS A PLANE. IT IS …

One morning, on a recent trip, as I settled down at an airport hotel to my morning glass of orange juice, a headline from the front page of that day's *The Wall street Journal* grabbed my attention. No, it was not about a Superman sighting over the New York skyline. Rather, *The WSJ* reporter Devlin Barrett was breaking news [1] of a Department of Justice project to scoop data from unsuspecting mobile phones by low-flying aircraft equipped with a device called a "dirtbox." The story was picked up rapidly by international news media, for example, *The Guardian* [2] in the United Kingdom and *Le Figaro* [3] in France. A one-line definition of the term dirtbox appeared on Wikipedia [4] with more "fast facts" about the gizmo listed on another website [5]. So what is this hullabaloo all about? And, what really is a dirtbox?

As *The WSJ* reported [1], this US Marshals Service Program, which has been functional since 2007, operates Cessna aircraft in a "high-tech hunt for criminal suspects" from a number of metropolitan airports, covering most of the populous regions of the country. These planes carry devices (dubbed "dirtboxes") that "mimic cell towers of large telecommunications firms [e.g., Verizon] and trick [mobile] telephones into reporting their unique registration information." The devices are apparently [1] about two square feet in size and are referred to as "dirtboxes" after the initials of the company DRT, Inc. [6], a subsidiary of Boeing Corp. Here is a *plausible* mode of operation of a "dirtbox" [5]:

> *"Mobile phones are programmed to connect with the closest signal tower...Boxes in planes could automatically assure mobile phones they are the optimal signal tower, then accept identifying information from handsets seeking connections. Fake cell towers could then pass connections onto real signal towers, remaining as a conduit with the ability to tune into or block digital transmissions. Hackers refer to such tactics as 'man-in-the-middle attacks."*

According to *The WSJ* report [1], the technology targets cellphones linked to fugitives and drug dealers but, in the process, also collects information on cellphones belonging to innocent people in the vicinity. Even having encryption on a mobile device such as the new iPhone 6 does not prevent a "dirtbox" from capturing its unique registration information. It is claimed [1] that the device determines which phones belong to potential criminals and "lets go" of the innocent phones.

A federal appeals court ruled earlier in 2014 (in a different scenario) that overcollection and stockpiling of data by investigators was a violation of the Constitution. While the Department of Justice refused to comment on this specific program on record, a departmental official, speaking on condition of anonymity, told *the Guardian* [2]:

> *"Discussion of sensitive law enforcement equipment and techniques would allow criminal defendants, criminal enterprises or foreign powers to determine our capabilities and limitations in this area. In deploying any such equipment or tactics our federal law enforcement agencies comply with federal law, including by seeking court approval."*

I doubt that would have relieved George Orwell's mind about Big Brother's intentions.

REFERENCES

[1] D. Barrett, "Airplanes Secretly Track U.S. Cellphones," *The Wall Street Journal*, pp. A1 and A6, November 14, 2014.

[2] "Government Planes Mimic Cellphone Towers to Collect User Data – Report," *The Guardian* [Online]. Available: http://www.theguardian.com/world/2014/nov/14/government-planes-mimic-cellphone-towers-to-collect-user-data-report (accessed January 5, 2016).

[3] "Des avions espions ont fouille les telephones des Americains," *Le Figaro* [Online]. Available: http://www.lefigaro.fr/secteur/high-tech/2014/11/14/01007-20141114ARTFIG00170-des-avions-espions-ont-fouille-les-telephones-des-americains.php (accessed January 5, 2016).

[4] "Dirtbox," *Wikipedia* [Online]. Available: http://en.wikipedia.org/wiki/Dirtbox (accessed January 5, 2016).

[5] "Dirtbox Devices: 5 Fast Facts You Need to Know" [Online]. Available: http://heavy.com/tech/2014/11/dirtbox-devices-spying-justice-department-boeing-fake-cell-tower/ (accessed January 5, 2016).

[6] Digital Receiver Technology, Inc. website [Online]. Available: http://www.drti.com/ (accessed January 5, 2016).

(The original version of the column appeared in "AP-S turnstile," *IEEE Antennas and Propagation Magazine*, vol. 56, no. 6, pp. 194–195, December 2014.)

NOTE

1. To learn more about how cell phone networks are configured, see for example: http://www.mat.ucsb.edu/~g.legrady/academic/courses/03w200a/projects/wireless/cell_technology.htm (accessed January 5, 2016).

DID YOU KNOW?

A FUN QUIZ (IV)

1. The route map of the London Underground (the "tube") is featured everywhere from tee shirts to coffee mugs. Recently, Japanese designer Yuri Suzuki used the tube schematic to make a printed circuit board and then, by installing components at strategic locations, turned it into a functioning _____.
    (a) GPS receiver
    (b) Mobile phone
    (c) Radio
    (d) None of the above

2. In a potential search-and-rescue breakthrough, researchers at the North Carolina State University have succeeded in creating a *wireless* electronic/biological interface, which allows them to steer *remotely* _____ in different directions.
    (a) Mice
    (b) Ants
    (c) Cockroaches
    (d) None of the above

3. The dream of zero standby leakage current may become a reality if researchers at the University of Minnesota succeed in replacing silicon transistors with nanometer dimension _____ in computational devices.
    (a) Light emitting diodes
    (b) Microelectromechanical (MEM) devices
    (c) Magnets
    (d) None of the above

4. Nanoantennas:
    (a) Convert light to electric power and vice versa
    (b) Transmit and receive "nanowaves"
    (c) Are antennas made of metamaterials
    (d) None of the above

From ER to E.T.: How Electromagnetic Technologies Are Changing Our Lives, First Edition. Rajeev Bansal.
© 2017 by The Institute of Electrical and Electronic Engineers, Inc. Published 2017 by John Wiley & Sons, Inc.

5. Residential "smart meters" used by electric utilities
   (a) Transmit radio signals
   (b) Receive radio signals
   (c) Transmit and receive radio signals
   (d) None of the above

6. Moving toward the holy grail of synthetic replacement organs, a team of researchers from Boston has successfully created a fine network of _____ based electronic sensors fully "embedded" in living tissue.
   (a) Silicon field effect transistor (FET)
   (b) Fiber optics
   (c) Microelectromechanical (MEM) device
   (d) None of the above

7. The proposed ground-based radio telescope *Square Kilometer Array* (SKA) will probe the early years after the Big Bang to look for answers to fundamental questions about our universe. It will be located in:
   (a) California
   (b) South Africa and Australia
   (c) Chile
   (d) None of the above

8. *Renal denervation* is a procedure which lowers blood pressure by using _____ energy to block the renal (kidney) nerves.
   (a) Optical
   (b) Infrared
   (c) Radiofrequency (RF)
   (d) None of the above

9. *Distant reading*, proposed by Franco Moretti (the founder of the Stanford Literary Lab), aims to
   (a) Understand literature by aggregating and analyzing massive amounts of literary data
   (b) Appreciate a non-western culture by reading its texts in the original language.
   (c) Make all literary texts universally available over the internet.
   (d) None of the above.

10. The *Energy Harvesting Forum*:
    (a) Supports subsidies for corn farmers for increased ethanol production
    (b) Focuses on techniques for capturing small amounts of energy from the environment for use by low-power devices
    (c) Encourages conversion from fossil fuels to alternate energy sources
    (d) None of the above

ANSWERS

1. **(c)** Radio
   Source: M. Prigg, "Sound of the Underground: Iconic London Tube Map Recreated as a Working Radio" [Online]. Available: http://www.dailymail.co.uk/sciencetech/article-2208390/Iconic-London-map-recreated-working-radio.html (accessed January 5, 2016).

2. **(c)** Cockroaches
   Source: "Cockroaches Controlled by Remote Control: A Search-and-Rescue Breakthrough?" [Online]. Available: http://theweek.com/article/index/233087/cockroaches-controlled-by-remote-control-a-search-and-rescue-breakthrough (accessed January 5, 2016).

3. **(c)** Magnets
   Source: "Prof. Kim pushes the envelope for chip speed, efficiency, and reliability," *Signals*, ECE Department Newsletter, University of Minnesota, Fall 2012.

4. **(a)** Convert light to electric power and vice versa
   Source: "Survival of the Fittest Nanoantenna" [Online]. Available: http://physicsworld.com/cws/article/news/2012/sep/28/survival-of-the-fittest-nanoantenna (accessed January 5, 2016).

5. **(c)** Transmit and receive radio signals
   Source: "How Smart Meters Work" [Online]. Available: https://www.bge.com/smartenergy/smartgrid/smartmeters/Pages/How-Smart-Meters-Work.aspx (accessed January 5, 2016).

6. **(a)** Silicon field effect transistor (FET)
   Source: "Living Tissue is Laced With Electronic Sensors" [Online]. Available: http://physicsworld.com/cws/article/news/2012/aug/29/living-tissue-is-laced-with-electronic-sensors (accessed January 5, 2016).

7. **(b)** South Africa and Australia
   Source: "Square Kilometre Array" [Online]. Available: http://www.skatelescope.org/ (accessed January 5, 2016).

8. **(c)** Radiofrequency (RF)
   Source: "Renal Denervation Achieves Significant and Sustained Blood Pressure Reduction, Study Suggests" [Online]. Available: http://www.sciencedaily.com/releases/2012/08/120827074032.htm (accessed January 5, 2016).

9. **(a)** Understand literature by aggregating and analyzing massive amounts of literary data
   Source: K. Schultz, "Distant Reading," *The New York Times Book Review*, p. 14, Sunday, June 26, 2011.

10. (b) Focuses on techniques for capturing small amounts of energy from the environment for use by low-power devices
Source: "Energy Harvesting Forum" [Online]. Available: http://www.energy harvesting.net/ (accessed January 5, 2016).

(The original version of the quiz appeared in "AP-S turnstile," *IEEE Antennas and Propagation Magazine*, vol. 55, no. 1, pp. 170–174, February 2013.)

CHAPTER 5

HEALTH EFFECTS OF ELECTROMAGNETIC FIELDS

"It's dangerous… like the mediaeval inquisitors who demanded that heretics prove their innocence. You cannot always prove your innocence."

—Jean de Kervasdoué (1944–)

5.1 SAY AU REVOIR TO CELL PHONES?

From ER to E.T.: How Electromagnetic Technologies Are Changing Our Lives, First Edition. Rajeev Bansal.
© 2017 by The Institute of Electrical and Electronic Engineers, Inc. Published 2017 by John Wiley & Sons, Inc.

Europeans are generally thought of as being big fans of everything "hi-tech," from their envied high-speed railway networks to the (now sadly defunct) supersonic Concorde. I recall being regularly surprised, during a sabbatical in the United Kingdom in 1994, by the ubiquity of cell phones in Europe. It seems commonplace now in the United States, but back then the sight of a young neighbor in Oxford, who would step out of his apartment with the cell phone glued to his ear, lock the door, and step into a waiting taxi, all without missing a clipped syllable of the conversation, made me feel like a country cousin visiting the metropolis. Therefore, a recent blog [1] by the Paris-based correspondent for the *Times* (London) seemed a bit paradoxical.

It appears that in France, where the bulk of electricity is produced by 59 nuclear reactors without the population going ballistic about the fears of another Chernobyl, both the government and the people have been lately displaying a lot of skittishness about the radiation from cell phones. A university library in Paris, fearing for the health of its patrons, at one point did away with its wireless network [2]. There were reports [1, 3] that cell phones were to be banned from primary schools and operators were being told to offer handsets that allow only text messages.

The next front in this French battle for limiting exposure to electromagnetic radiation is cell phone towers. Both the French government and the cell phone operators are being attacked by numerous local and national groups demanding that cell phone towers be removed from locations close to schools, hospitals, and homes. One such national group calls itself "Robin-des-Toits," or Robin of the Roofs, which is a pun on Robin des Bois, or Robin of the Woods (Robin Hood) [1]. The website [4] of the group posts extensive "evidence" on the putative dangers of electromagnetic radiation.

Such campaigns are neither new nor unique to France but they have enjoyed much greater success there in legal proceedings. For example, an appeals court in Versailles ordered the French operator Bouygues to take down a tower near Lyons, because local families were concerned about health effects. As Bremner [1] reported: "The judges agreed that there was no evidence of a threat, but they said there was no guarantee that a risk did not exist. The 'feeling of anxiety' of the inhabitants was therefore justified."

In a particularly interesting case [5], family members described complaints ranging from a metallic taste in the mouth to nosebleeds, all attributed to a cell phone tower recently installed across from its apartment building. The family even covered the apartment windows with aluminum foil and other "protective filters" to ward off the ill effects of the radiation from the tower. For its part, the operator of the cell phone tower (Orange) dryly noted that the electronic bay for the tower had not been installed yet and, therefore, the tower was not even active.

In an interview with *le Journal du Dimanche*, Jean de Kervasdoué, a former national director of French hospitals, made an important point on the futile attempts to seek zero risk when he observed [1]: *"It's dangerous… like the mediaeval inquisitors who demanded that heretics prove their innocence. You cannot always prove your innocence."*

REFERENCES

[1] C. Bremner's blog for the *Times* (London) is available (subscription) at: http://www. thetimes.co.uk/tto/public/profile/Charles-Bremner (accessed August 27, 2015).

[2] The Wi-Fi ban at the Paris-III university was reported by *Le Monde* in its May 13, 2009 edition.

[3] A television report on cell phone ban in French schools was broadcast by WKYC.

[4] The website of the French advocacy group "Robin-des-Toits" is at: http://www. robindestoits.org/ (accessed August 27, 2015).

[5] A report on complaints about an "inactive" cell phone tower is located at: http://www. bestofmicro.com/actualite/26785-antenne-relais.html (accessed August 27, 2015).

(The original version of the column appeared in "AP-S turnstile," *IEEE Antennas and Propagation Magazine*, vol. 51, no. 3, p. 152, June 2009.)

NOTES

1. Public concerns about the siting of cell phone towers are discussed also in Section 5.3.

2. Current research results on the safety of cell phones are posted on the website of the US Food and Drug Administration (FDA): http://www.fda.gov/Radiation-EmittingProducts/ RadiationEmittingProductsandProcedures/HomeBusinessandEntertainment/CellPhones/ ucm116335.htm (accessed April 30, 2016).

3. 🖼 Which of the following emits microwave radiation?
   (a) The sun
   (b) The human body
   (c) Police radar
   (d) All of the above

   > 🖼 (d) All of the above
   > The power density from the sun is 100 mW/cm^2 on a clear day. However, in the microwave range (2–100 GHz), it is roughly 80 dB below the current 1 mW/cm^2 safety limit prescribed by the current IEEE standard C95.1-2005. The typical exposure level to police radar is 1 microwatt/cm^2. And, yes, even the human body *emits* (as part of the tail of its black-body radiation) microwave radiation at a level of about 0.3 mW/cm^2.
   > *Source:* R. Bansal, "Pop quiz: EMF and your health," *IEEE Potentials*, pp. 3–4, August/September 1997.

4. Concerns about the putative health hazards of wireless networks in public spaces continue to pop up in European towns. For example, here is a 2016 story about Wi-Fi in schools in the small town of Borgofranco d'Ivrea in the Piedmont region of Italy: http://www.thelocal.it/ 20160108/italy-town-turns-off-school-wifi-over-health-concerns (accessed April 30, 2016).

5.2 ELECTROMAGNETIC HYPERSENSITIVITY

For some years now, I have been serving as a member of the IEEE Committee on Man and Radiation (COMAR). Let me briefly recap the committee's modus operandi: COMAR critically examines and interprets the literature on biological effects. Its findings are usually reported in the form of *technical information statements* (TISs). Before being made public, these reports are thoroughly reviewed by the Committee members, and upon approval represent the consensus opinion. They are usually published in the *IEEE EMBS Magazine* and posted on the COMAR website [1]. It should be emphasized that COMAR does not establish safety standards, but it has an interest in the standards activity within its scope.

A TIS approved by the COMAR membership deals with the topic of **Electromagnetic Hypersensitivity**. The full text is available both in print form [2] and on the web [1]. I have summarized the main points below.

What is Electromagnetic Hypersensitivity(EHS)?

Some people experience various health symptoms that they attribute to exposure to electric and magnetic fields from sources ranging from consumer electronics equipment (TVs, computers, etc.) in stores [3] to household appliances and cellular phones. This *perceived* sensitivity to electromagnetic fields is usually reported in situations where the fields are known to be well below the safety thresholds incorporated in various international standards. In certain cases, the experience is severe enough for individuals to quit work or make radical adjustments to their lifestyle, for example, sleeping under metal blankets. In Sweden, a group called the Swedish Association for the ElectroSensitive (FEB) provides support and advocacy [4].

What Are the Main Health Symptoms?

According to a 1997 report (English summary is available at Reference [5]) prepared for the European Commission, the commonly experienced symptoms include:

- Nervous system symptoms, for example, fatigue, stress, sleeping problems
- Skin symptoms including various sensations and rashes
- Pains and aches

Depending upon the methodology used, Swedish surveys about the prevalence of the symptoms resulted in estimates ranging from a few individuals per million (too low?) to a few tenths of a percent of the general population (too high?). EHS has a higher prevalence in Sweden, Germany, and Denmark than in France, United Kingdom, and Austria.

What Do Provocation Studies Reveal About EHS?

In provocation studies, electromagnetically hypersensitive individuals are exposed to electric and magnetic fields under controlled laboratory situations with a view to exploring the link between the fields and the symptoms exhibited. Taken as a whole, the laboratory studies to date have not been able to link actual field exposures to EHS symptoms experienced. Electromagnetically hypersensitive people do not appear to be any better at detecting the presence of EM fields than the general population.

Are There Other Disorders Like EHS?

EHS has been compared with multiple chemical sensitivities (MCS). In both cases, the symptoms may have a variety of causes, the exposure levels to chemicals or EM fields are considerably lower than commonly accepted thresholds for adverse effects, provocation studies are unsuccessful in establishing a connection between the exposure and the symptoms, and they remain poorly understood. Investigators have also noted similarities between EHS and "microwave illness," which has been reported in the Russian and Eastern European literature as case reports (without corroborating epidemiological data) since the 1970s.

How Can EHS Sufferers Be Helped?

It needs to be emphasized that EHS is a real problem for the affected person. The recommended strategy to help an EHS sufferer is to focus on the health symptoms of the person and examine and mitigate potentially contributing environmental/occupational factors (e.g., lighting, air quality, ergonomic issues). This is best accomplished in cooperation with a physician, hygienist, and, where appropriate, a psychotherapist. The patient should also be provided with scientifically based information about safety issues as they relate to electromagnetic fields [5].

REFERENCES

[1] COMAR website [Online]. Available: http://ewh.ieee.org/soc/embs/comar/ (accessed August 27, 2015).

[2] M. C. Ziskin, "Electromagnetic hypersensitivity-A COMAR technical information statement," *IEEE Engineering in Medicine and Biology Magazine*, vol. 21, no. 5, pp. 173–175, September/October 2002.

[3] Published in the "To Your Health" column in the *Austin Chronicle*, January 10, 2003.

[4] FEB website [Online]. Available: http://www.feb.se/FEB/index.html (accessed August 27, 2015).

[5] U. Bergqvist and E. Vogel (eds.). "Possible health implications of subjective symptoms and electromagnetic fields." A report prepared by a European group of experts for the European Commission, DGV. Arbete och Hälsa, 1997:19. Swedish National Institute for Working Life, Stockholm, Sweden. ISBN 91-7045-438-8. (In Swedish).

(The original version of the column appeared in "AP-S turnstile," *IEEE Antennas and Propagation Magazine*, vol. 45, no. 2, pp. 102–103, April 2003.)

NOTE

1. Electromagnetic hypersensitivity is also discussed in Section 5.4.

5.3 FROM BELL TOWER TO CELL TOWER

Anyone who has driven around rural New England can relate to the sight of white church steeples piercing the rolling green landscape at regular intervals. While these steeples generally house nothing more exotic than church bells, cellular service providers in many areas have been trying to subject these towers to multi-tasking by installing base stations within. Some time ago, I received an email from the president of the board of directors of a local nursery school. The nursery school rents some rooms on the premises of a church, which is considering leasing space in the steeple to a cellular service provider. Since some of the parents expressed concerns about exposure to RF/microwave radiation from the proposed base station, the president, who found out about my interest in biological effects of microwave radiation through one of my colleagues, invited me to a meeting of the school board.

In preparation for the meeting, I went through my files and surfed the web for recent stories on safety issues associated with base stations. Since the subject may be of interest to many readers for personal or professional reasons, I have presented below some talking points (suitable for a non-technical audience) with links to the source material.

Base Stations

Each base station is a low-power radio station (generally operating in the 800–1000 MHz or 1850–1990 MHz frequency range) and serving the customers within a small geographic region called a "cell." The location of a base station is governed by two needs of the cellular service provider: providing adequate coverage throughout the entire service area and providing enough capacity (free channels) to accommodate a growing customer base. As a system grows, the cell size may shrink to increase capacity but the radiated power level also goes down to avoid interference among adjacent cells [1].

Antennas

While some base stations may employ pole-like "omnidirectional" antennas (transmitting uniformly in the horizontal plane), others use three groups of "sector" antennas (oriented 120 degrees apart), which are rectangular panels (about 1 × 4 feet in dimension) [1–3].

Radiated Power

The Federal Communications Commission (FCC) permits an effective radiated power (ERP) up to 500 watts per channel. A cellular base station may generally transmit using 21 channels per sector. However, it is unlikely that all the transmitters would be operating simultaneously [1–2].

Ground Level Radiation

At ground level, the power density is relatively low near the base of the tower, since the main beam from the antenna is directed toward the horizon. Generally speaking, the signal strength on the ground reaches a maximum 50–200 m from the base of the tower, and then decreases as one moves beyond that distance [1].

Exposure Limits

At all places on the ground, the signal strength is well below the regulatory guidelines established by the FCC (as well as those endorsed by major international organizations). Typical exposure levels are 100 times or more below these limits. Also, exposure to persons inside a building with a roof-mounted base station antenna is invariably very low. As Professor Ken Foster noted in his testimony [4] in a zoning board hearing in Pennsylvania, people would only risk being exposed over limits if they were a few feet away from the antenna outside of the top of the actual steeple [1, 2, 4].

Scientific Basis of Regulatory Guidelines

Regulatory standards are derived through a careful, open consensus process "to identify the lowest exposure level for which *credible* evidence exists for any *reproducible* adverse effect that can be related to human health" [emphasis added] [1, 5]. Exposure guidelines incorporate significant (10–50) safety factors to keep exposures well below hazardous levels. The situation was summarized well by Dr. Jessica Leighton, the then Assistant Commissioner for the Bureau of Environmental Disease Prevention, NYC Department of Health and Mental Hygiene, in her testimony [5] before the City Council on the subject of "Installation of Cell Phone Antennas and Base Stations on Rooftops throughout New York City":

> *"Current epidemiological and clinical evidence does not support findings of an association between radio frequencies and adverse health effects such as cancer, reproductive effects, neurological effects or others…. The principal means by which cell phone radio frequencies are understood to affect cells and animals is through the heating of tissue. Current FCC regulations are intended to protect against these effects by limiting the power of individual and clustered antenna installations."*

Legal Position

According to Section 704 of the Telecommunications Act of 1996, "no State or local government or instrumentality thereof may regulate the placement, construction, and modification of personal wireless service facilities on the basis of the environmental effects of radio frequency emissions to the extent that such facilities comply with the Commission's regulations concerning such emissions [3]."

REFERENCES

[1] "Safety Issues Associated With Base Stations Used for personal Wire Communications," Technical Information Statement (TIS) prepared by COMAR, September 2000. Available: http://www.ewh.ieee.org/soc/embs/comar/base.htm (accessed August 27, 2015).

[2] The communication "Are Cellular and PCS Towers and Antennas Safe," August 10, 2001, was prepared by the University of California, Irvine.

[3] "Cell Phones," available at the FDA website at http://www.fda.gov/radiation-emitting products/radiationemittingproductsandprocedures/homebusinessandentertainment/cell phones/default.htm (accessed April 30, 2016).

[4] Professor Ken Foster provided testimony before a zoning board hearing in Pennsylvania.

[5] Dr. Jessica Leighton provided testimony on April 30, 2004 before the City Council on the subject of "Installation of Cell Phone Antennas and Base Stations on Rooftops throughout New York City."

(The original version of the column appeared in "Microwave surfing," *IEEE Microwave Magazine*, vol. 5, no. 3, pp. 30–32, September 2004.)

NOTES

1. Public concerns about the siting of cell phone towers are discussed also in Section 5.1.
2. To learn more about how cell phone networks are configured, see for example: http://www. mat.ucsb.edu/~g.legrady/academic/courses/03w200a/projects/wireless/cell_technology. htm (accessed April 30, 2016).
3. Textbook resources:
   (i) W. H. Hayt and J. A. Buck, *Engineering Electromagnetics*, 8th ed., McGraw-Hill, New York, 2012. The basics of radiation and antennas are presented in Chapter 14.
   (ii) F. T. Ulaby and U. Ravaioli, *Fundamentals of Applied Electromagnetics*, 7th ed., Prentice Hall, Upper Saddle River, NJ, 2015. The basics of radiation and antennas are presented in Chapter 9.

5.4 NOCEBO: READING THIS COLUMN MAY AFFECT YOUR HEALTH

In Reference [1], I mentioned the following interesting case from France:

> *"Family members described complaints ranging from a metallic taste in the mouth to nosebleeds, all attributed to a cell phone tower recently installed across from its apartment building. The family even covered the apartment windows with aluminum foil and other 'protective filters' to ward off the ill effects of the radiation from the tower. For its part, the operator of the cell phone tower (Orange) dryly noted that the electronic bay for the tower had not been installed yet and, therefore, the tower was not even active."*

More recently, Professor Lin discussed the case of hypersensitivity to electromagnetic fields [2]:

> *"The syndrome of electromagnetic hypersensitivity (EHS) consists of nervous-system symptoms, such as headache and fatigue; skin symptoms, such as facial irritations and rashes; as well as other nonspecific health-related symptoms. One of the most renowned cases of EHS is the reported hypersensitivity of Gro Harlem Brundtland, the former Prime Minister of Norway. She was the Director-General of the World Health Organization (WHO) from 1998 to 2003. She had not publicly talked about her EHS*

for more than 10 years. However, that ended when she told a reporter, 'I avoid talking on mobile phone,' in response to a newspaper article alleging that she now uses a mobile phone, according to a former top aide at WHO, Jonas Gahr Støre, the current Norwegian Minister of Health."

Professor Lin goes on to say: "The reported evidence suggests that while the phenomenon of hypersensitivity may be real, the questions as to whether the symptoms are associated with cell-phone use or how best to study EHS in a controlled laboratory investigation remain controversial."

Recently, I received a copy of an interesting paper from Ric Tell, Chair of COMAR [3], an IEEE/EMBS committee, of which I am a member. The paper [4] by Michael Witthoeft and G. Rubin asks the provocative question "Are media warnings about the adverse health effects of modern life self-fulfilling?" and proceeds to investigate it experimentally in the context of electromagnetic hypersensitivity, for which the authors use the term idiopathic environmental intolerance attributed to electromagnetic fields (IEI-EMF). In this study, "participants (N = 147) were randomly assigned to watch a television report about the adverse health effects of WiFi (n = 76) or a control film (n = 71). After watching their film, participants received a sham exposure to a WiFi signal (15 min). The principal outcome measure was symptom reports following the sham exposure. Secondary outcomes included worries about the health effects of EMF, attributing symptoms to the sham exposure and increases in perceived sensitivity to EMF." The authors found that watching the television report about the adverse health effects of Wi-Fi increased EMF-related worries, postsham-exposure symptoms among participants with high pre-existing anxiety, the likelihood of symptoms being attributed to the sham exposure among people with high anxiety, and the likelihood of people who attributed their symptoms to the sham exposure believing themselves to be sensitive to EMF. This is the so-called Nocebo effect [5] in play:

"Nocebo" (meaning "I shall harm" in Latin) is the dastardly sibling of placebo ("I shall please"). In a placebo response, a sham medication or procedure has a beneficial health effect as a result of a patient's expectation. Sugar pills, for example, can powerfully improve depression when the patient believes them to be antidepressants. But, researchers are learning, the reverse phenomenon is also common: negative expectations can actually cause harm."

Witthoeft and Rubin conclude: *"Media reports about the adverse effects of supposedly hazardous substances can increase the likelihood of experiencing symptoms following sham exposure and developing an apparent sensitivity to it."* Their recommendation: *"Greater engagement between journalists and scientists is required to counter these negative effects."*

If you have read this far and feel dizzy, you may blame the column for it. Sorry, mea culpa.

REFERENCES

[1] R. Bansal, "AP-S turnstile: say au revoir to cell phones," *IEEE Antennas and Propagation Magazine*, vol. 51, no. 3, p. 152, June 2009.

[2] J. Lin, "Telecommunication health and safety: the case of hypersensitivity to electromagnetic fields," *IEEE Antennas and Propagation Magazine*, vol. 55, no. 4, pp. 258–260, August 2013.

[3] COMAR website [Online]. Available: http://ewh.ieee.org/soc/embs/comar/ (accessed January 6, 2016).

[4] M. Witthoeft and G. Rubin, "Are media warnings about the adverse health effects of modern life self-fulfilling? An experimental study on idiopathic environmental intolerance attributed to electromagnetic fields (IEI-EMF)," *Journal of Psychosomatic Research*, vol. 74, no. 3, pp. 206–212, March 2013.

[5] M. Scedallari. "Worried Sick," *The Scientist* [Online]. Available: http://www.the-scientist.com/?articles.view/articleNo/36126/title/Worried-Sick/ (accessed August 23, 2016).

(The original version of the column appeared in "AP-S turnstile," *IEEE Antennas and Propagation Magazine*, vol. 55, no. 6, pp. 178–179, December 2013.)

NOTES

1. Reference [1] is included in this book as Section 5.1.
2. Section 5.2 discusses electromagnetic hypersensitivity in more detail.

5.5 MAGNETIC PULL: BIOLOGICAL EFFECTS OR MEDICAL APPLICATIONS?

One of the questions on the IMS 2005 Quiz (*IEEE Microwave Magazine*), related to the magnetic fields used in magnetic resonance imaging (MRI) machines. In particular, it asked:

An MRI machine uses …

(a) High-intensity *static* magnetic fields
(b) *Time-varying* magnetic fields in the kHz range

(c) Radio-frequency (RF) fields in the MHz range

(d) All of the above

The correct answer, (d) All of the above, highlights the fact that an MRI machine employs all three types of fields in order to accomplish three-dimensional imaging for medical diagnostics. Therefore, in terms of safety, electromagnetic field exposure to patients and medical personnel needs to be assessed for all three types of fields. This is a much greater concern for *open* MRI systems, where the magnet surrounds the patient only partially in order to allow the medical personnel to perform procedures on the patient, guided by the MRI images.

The largest of the three types of the MRI magnetic fields is the *static* magnetic field. These intense static magnetic fields are generally considered safe for the patients and the attending personnel (the present IEEE exposure standards do not even specify a limit for exposure to *static* magnetic fields). However, as a story in *The New York Times* [1] illustrates, they can pose a serious accidental hazard, when manufacturer-specified precautions are not taken. The powerful MRI magnets can pull all kinds of ferromagnetic objects (e.g., a revolver, a wheelchair, an acetylene tank) from the hands of careless workers into the scanner. A website [2] maintained by Dr. Moriel Ness Aiver, who taught MRI safety, has photographs of many unusual objects that flew into and were lodged in MRI machines. The US FDA also maintains a record of MRI accidents, but that database is drawn primarily from reports filed by MRI manufacturers, who find out about these accidents only when the machine is damaged by the accident. However, the FDA website [3] on MRI safety cites the case of a 6-year-old boy who was killed, while undergoing an MRI scan, by a ferromagnetic oxygen tank that was pulled into the MRI scanner by the strong magnetic field. Fortunately, most modern surgical implants and devices (staples, clips, artificial joints, pacemakers, etc.) are made of non-ferromagnetic materials such as titanium or stainless steel, but there have been accidents with patients who had older metal clips or pacemakers implanted in their bodies.

There are well-established standards that address MRI safety. For example, the FDA website cites the following:

- ASTM F2052-00 Standard Test Method for Measurement of Magnetically Induced Displacement Force on Passive Implants in the Magnetic Resonance Environment
- ASTM F2119-01 Standard Test Method for Evaluation of MR Image Artifacts From Passive Implants
- IEC 601-2-33—Medical Electrical Equipment—Part 2: Particular requirements for the safety of magnetic resonance equipment for medical diagnosis

The problem is that most accidents are caused by human error and while the FDA approves the MRI scanners as medical devices, it does not regulate how the users operate the machines. Scanner manufacturers can recommend safer room designs, for example, locked doors, offer safety training and advice, but they cannot legally enforce them. As the number of scanners has multiplied to more than 10,000 units in the United States and their magnets have become more potent, the risk of careless accidents has also increased. While a new generation of detectors that can screen for ferromagnetic materials is available, meticulous safety training remains crucial to accident-free operation.

REFERENCES

[1] D. McNeil, Jr., "M.R.I.'s Strong Magnets Cited in Accidents," *The New York Times*, August 19, 2005.

[2] http://www.simplyphysics.com/flying_objects.html (accessed April 30, 2016).

[3] Current US FDA website on MRI [Online]. Available: http://www.fda.gov/Radiation-EmittingProducts/RadiationEmittingProductsandProcedures/MedicalImaging/MRI/default.htm (accessed April 30, 2016).

(The original version of the column appeared in "Microwave surfing," *IEEE Microwave Magazine*, vol. 6, no. 4, pp. 54–56, December 2005.)

NOTES

1. To learn more about the operation of an MRI machine, see for example: http://science.howstuffworks.com/mri.htm (accessed April 30, 2016).

2. The COMAR technical information statement (TIS) on the exposure of medical personnel from open MRI systems is available through the COMAR website: http://ewh.ieee.org/soc/embs/comar/ (accessed April 30, 2016).

3. Which produces the *largest* magnetic field?

   (a) An MRI machine

   (b) The earth

   (c) A 13 kV distribution line along the street

   (d) A 735 kV transmission line from a power generating station

   (a) An MRI machine

   MRI machines use very large static magnetic fields (more than ten thousand gauss) for diagnostic purposes. Among the remaining choices, the earth's *static* (dc) magnetic flux density in air (about half a gauss) is roughly 200 times larger than the ac magnetic flux density from typical 60 Hz distribution lines. Lastly, since the magnetic field varies with the line *current*, a 13 kV distribution line may (in some situations) generate a larger magnetic field than a 735 kV transmission line [1 T (tesla) = 1 Wb/m^2 = 10,000 G (gauss)]. *Source*: R. Bansal, "Pop quiz: EMF and your health," *IEEE Potentials*, pp. 3–4, August/September 1997.

4. Textbook resources:

   (i) W. H. Hayt and J. A. Buck, *Engineering Electromagnetics*, 8th ed., McGraw-Hill, New York, 2012. Magnetic forces are discussed in Chapter 8.

(ii) F. T. Ulaby and U. Ravaioli, *Fundamentals of Applied Electromagnetics*, 7th ed., Prentice Hall, Upper Saddle River, NJ, 2015. Magnetic forces are discussed in Chapter 5.

5.6 CLOSE ENCOUNTERS WITH RADIATION OF THE OTHER KIND

Let's start with a little quiz:

In the summer of 2001, Continental rerouted many of its New York to Hong Kong flights from their polar routes to lower latitudes over Canada ...

(a) *To avoid seasonal headwinds*

(b) *To ensure good radio communication*

(c) *To reduce potential radiation exposure*

(d) *All of the above*

If you chose (a), (b), or (c), you deserved partial credit; however, if you chose (d), you aced the test. As a story [1] in The *New York Times* pointed out, during its "storm" season, the sun emits proton bursts (solar flares), which can expose people flying high above the earth's protective atmosphere to *ionizing* radiation (in contrast with microwave radiation which is *non-ionizing*). The Times hastened to add: "There is no evidence so far that this exposure is dangerous. In fact, the flares' disruption of radio communication is a bigger threat to passengers."

While the bursts of ionizing radiation accompanying solar flares are infrequent events, frequent fliers are exposed to galactic cosmic radiation all the time. A primary source of this galactic cosmic radiation is believed to be exploding stars (supernovae). At aircraft flight altitudes, the galactic cosmic radiation consists mainly of neutrons, protons, electrons, x-rays, and gamma rays [2]. As a result of this "background" radiation, both the European Authorities and the Federal Aviation Administration (FAA) consider flight crews as radiation workers. Some airlines, including British Airways, reassign female crewmembers to ground duties as soon as they inform the management about their pregnancy. The Radiobiology Section of the FAA Office of Aerospace Medicine [2] offers the following advice: "With regard to occupational exposure to radiation during pregnancy, the pregnant crewmember and management should work together to ensure the radiation exposure of the conceptus does not exceed recommended limits." The Office also provides an online calculator CARI-6

[3], which can be used to calculate the effective dose of galactic cosmic radiation received by a crewmember flying an approximate great-circle route (the shortest distance) between two airports. A complicating factor is that the human fetus is believed to be most vulnerable to radiation exposure in the first few weeks after conception, probably before the pregnancy is known [1].

In estimating radiation-induced health risks for aircrews, the FAA uses dose–effect relationships recommended by national and international organizations recognized for their expertise in evaluating radiation effects in humans. However, the FAA recognizes that "there is much uncertainty in the estimates because much of the original data comes from studies on individuals exposed to radiation at higher doses and dose rates and generally of lower energy than the galactic cosmic radiation to which aircrews are exposed" [2].

Interestingly, the uncertainty about the dose–effect relationship of ionizing radiation is not limited to galactic cosmic radiation, but applies just as well to other natural sources such as radon and radioactive substances in soil, food, and water. In a report last year on ionizing radiation standards, the General Accounting Office (GAO) said: "The standards administered by E.P.A. and N.R.C. to protect the public from low-level exposure do not have a conclusive scientific basis, despite decades of research." Since nearly a third of all people get cancer anyway at some point, the difficulty has been to find evidence that low doses of radiation cause excess cancer (cases that would not have otherwise occurred). A linear dose–effect relationship model is widely used to estimate radiation risks, but some scientists feel that below a threshold radiation does not pose a health hazard. Some even theorize that low radiation doses may actually be beneficial as they may activate body's natural defense mechanisms against cancer [4].

The uncertainties surrounding the health implications of low-level ionizing radiation may seem puzzling considering that [4] "there are probably more studies on the harmful effects of [ionizing] radiation than for any other toxic or noxious agents in the environment." However, it helps put in perspective the difficulties inherent in this kind of risk assessment. In particular, it is easier now to understand the following cautious GAO assessment [5] of the safety of non-ionizing radiation from cell phones:

"The consensus of the Food and Drug Administration (FDA), the World Health Organization, and other major health agencies is that the research to date does not show radiofrequency energy emitted from mobile phones has harmful health effects, but there is not yet enough information to conclude that they pose no risk."

REFERENCES

[1] M. Wald, "The Frequent Flier and Radiation Risk," *The New York Times*, June 12, 2001.

[2] The FAA Office of Aerospace Medicine: Radiobiology Research Team website [Online]. Available: http://www.faa.gov/data_research/research/med_humanfacs/aeromedical/radiobiology/ (accessed April 30, 2016).

[3] Galactic Radiation Received in Flight: Calculator. Available: http://jag.cami.jccbi.gov./cariprofile.asp (accessed April 30, 2016).

[4] G. Kolata, "For Radiation, How Much Is Too Much," *The New York Times*, November 27, 2001.

[5] The GAO report GAO-01-545 (05/07/2001) on mobile phone safety is available at: http://www.gao.gov/products/GAO-01-545 (accessed April 30, 2016).

(The original version of the column appeared in "Microwave surfing," *IEEE Microwave Magazine*, vol. 3, no. 4, pp. 44–46, March 2002.)

NOTE

1. Which of the following devices produces ionizing radiation?

   (a) A microwave oven

   (b) A TV transmission tower

   (c) A power substation transformer

   (d) None of the above

   (d) None of the above

   The electromagnetic spectrum can be roughly divided into two broad classes: ionizing (ultraviolet and higher frequencies including x- rays and gamma rays) and *non-ionizing* (visible light and lower frequencies including infrared, microwaves, and power frequencies). This division comes about because electromagnetic energy of a particular wavelength λ (or equivalently frequency $f = c/\lambda$) is associated with a specific photon energy E given by

   $$E(\text{electron volts}) = 1.24 \times 10^{-6}/\lambda(\text{meters}).$$

   If the photon energy exceeds approximately 10 eV (corresponding to a wavelength of 0.124 micrometer, which falls in the UV region), the photon can ionize materials, that is, break molecular bonds. Such ionization, particularly if it involves DNA molecules (which store genetic information and are located in the nuclei of tissue cells), can result in serious irreversible damage.

   In contrast, the energy produced by a 1 GHz photon is a mere fraction (one six-thousandth) of the thermal kinetic energy of a tissue molecule. This prevents it from breaking even the weakest bonds.

   Source: R. Bansal, "Pop quiz: EMF and your health," *IEEE Potentials*, pp. 3–4, August/September 1997.

DID YOU KNOW?

1. "Surrounded by Waves" is:
   (a) A popular satellite radio program
   (b) A music CD to soothe you to sleep
   (c) A documentary film about the putative hazards of wireless technologies
   (d) None of the above

2. According to a recent report released by the United Kingdom's Research Information Network, the Royal Astronomical Society, and the Institute of Physics, *Google Scholar* is used (to search for new research findings) by:
   (a) Roughly equal fractions of nanoscientists and astrophysicists
   (b) A much greater fraction of astrophysicists than that of nanoscientists
   (c) A much greater fraction of nanoscientists than that of astrophysicists
   (d) None of the above

3. The modern concept of "cellular communication," that is, breaking areas into hexagonal cells with antennas in the center, was proposed by Ring and Young of Bell labs in:
   (a) 1967
   (b) 1957
   (c) 1947
   (d) None of the above

4. An active electronically scanned array (AESA), when acting as a normal radar, broadcasts its microwave radiation over a wide area. At the touch of a button, however, the energy can be focused in order to:
   (a) Zap an incoming missile or aircraft
   (b) Track multiple warheads with exceptional accuracy
   (c) Detonate a buried land mine
   (d) None of the above

5. According to theoretical work by Professor Igor Smolyaninov of the University of Maryland, extremely high magnetic fields (think primordial universe) should make the vacuum behave like:

From ER to E.T.: How Electromagnetic Technologies Are Changing Our Lives, First Edition. Rajeev Bansal.
© 2017 by The Institute of Electrical and Electronic Engineers, Inc. Published 2017 by John Wiley & Sons, Inc.

(a) A conductor

(b) A semiconductor

(c) A metamaterial

(d) None of the above

6. The *Electric Cinderella Shoe*, when activated by a wireless necklace, can:

   (a) Deliver a 100,000 V jolt for neutralizing any human threat

   (b) Activate miniature electric motors to propel the wearer out of harm's way

   (c) Create a protective electric-field "bubble" around the wearer

   (d) None of the above

7. Based at NASA, SkyTran is an experimental:

   (a) Rocket for personal transportation

   (b) Low-cost, high-speed, magnetic-levitation-based *personal* transportation system

   (c) Replacement for the now-retired Space Shuttle

   (d) None of the above.

8. *Chaos Radar* uses signals generated by a "chaotic oscillator" to:

   (a) Confuse incoming missiles via electronic countermeasures

   (b) Confuse police radar units

   (c) See through walls

   (d) None of the above

9. *Tau Day*, celebrated by its proponents on 28 June, aims to dislodge the mathematical constant pi by arguably more useful tau, which equals:

   (a) 2 pi

   (b) 4 pi

   (c) 1/(4 pi)

   (d) None of the above

10. *Li-Fi* technology uses:

   (a) Lithium-ion batteries for powering autonomous wireless sensor networks

   (b) Rapid pulses of light for wireless communication

   (c) A hybrid of Wi-Fi and fiber optics to achieve an ultrahigh-bandwidth

   (d) None of the above

ANSWERS

1. (c) A documentary film about the putative hazards of wireless technologies
Source: "Surrounded by waves" [Online]. Available: http://icarusfilms.com/new2010/wav.html (accessed April 30, 2016).

2. **(c)** A much greater fraction of nanoscientists than that of astrophysicists
   Source: M. Durrani. "Online tools are 'distraction' for science" [Online]. Available: http://physicsworld.com/cws/article/news/48446 (accessed April 30, 2016).

3. **(c)** 1947
   Source: J. Browne, "Bracing for the cellular explosion," *Microwaves & RF*, p. 56, August 2011.

4. **(a)** Zap an incoming missile or aircraft
   Source: "Frying Tonight," *The Economist*, p. 89, October 15, 2011.

5. **(c)** A metamaterial
   Source: T. Wogan, "Was a Metamaterial Lurking in the Primordial Universe?" [Online]. Available: http://physicsworld.com/cws/article/news/48238 (accessed April 30, 2016).

6. **(a)** Deliver a 100,000 V jolt for neutralizing any human threat
   Source: D. Vye, R. Mumford, and P. Hindle, "The spy who loved microwaves," *Microwave Journal,* pp. 22–34, October 2011

7. **(b)** Low-cost, high-speed, magnetic-levitation-based *personal* transportation system
   Source: SkyTran [Online]. Available: http://www.skytran.us/ (accessed April 30, 2016).

8. **(c)** See through walls
   Source: D. Hambling, "Chaos Radar Uses Messy Signal to See Through Walls" [Online]. Available: http://www.newscientist.com/article/mg21128225.200-chaos-radar-uses-messy-signals-to-see-through-walls.html (accessed April 30, 2016).

9. **(a)** 2 pi
   Source: J. Palmer, "'Tau Day' Marked by Opponents of Maths Constant Pi" [Online]. Available: http://www.bbc.co.uk/news/science-environment-13906169 (accessed April 30, 2016).

10. **(b)** Rapid pulses of light for wireless communication
    Source: J. Condliffe, "Will Li-Fi Be the New Wi-Fi?" [Online]. Available: https://www.newscientist.com/article/mg21128225-400-will-li-fi-be-the-new-wi-fi/ (accessed April 30, 2016).

(The original version of the quiz appeared in "AP-S turnstile," *IEEE Antennas and Propagation Magazine*, vol. 54, no. 1, pp. 150–157, February 2012.)

CHAPTER 6

BIOMEDICAL APPLICATIONS

"Biology undergoes these revolutionary waves from time to time, after which nothing is ever the same. This is one of those times."

—Eric Davidson (1937–2015)

6.1 HOW MANY BIOLOGISTS DOES IT TAKE TO FIX A RADIO?

As a member of MTT-10, the IEEE technical coordinating committee on "Biological Effects and Medical Applications," I find it natural to take occasional peeks into the world of biology. 2003 was a particularly auspicious year for doing so, since it marked the golden anniversary of the discovery of the double-helix structure of the

From ER to E.T.: How Electromagnetic Technologies Are Changing Our Lives, First Edition. Rajeev Bansal.
© 2017 by The Institute of Electrical and Electronic Engineers, Inc. Published 2017 by John Wiley & Sons, Inc.

DNA molecule. That monumental find by Crick and Watson led to the astounding progress we have seen in the field of molecular biology, culminating recently in the sequencing of the human genome [1]. As Thomas Kuhn noted in his 1962 classic *The Structure of Scientific Revolutions*, the bulk of science proceeds within a given framework (a "paradigm") for certain historical reasons. The Crick–Watson discovery injected into biology a *reductionist* paradigm, whereby subsequent scientific progress was gauged by [1] "describing the smallest bits possible, usually one at a time—one stretch of DNA, one RNA, one protein."

How does the future of biology look? As a 2003 cover story [2] in *Time* put it, "Cracking the DNA code has changed how we live, heal, eat and imagine the future." However, not all biologists are as sanguine about the usefulness of the prevailing reductionist paradigm for the twenty-first century. Yuri Lazebnik, a molecular biologist himself, offers a contrarian view in a letter [3] he wrote to *Cancer Cell*. There he explores the potential pitfalls inherent in the current reductionist approach used in biology, by applying it to the problem of fixing a broken transistor radio, a reasonably complex but well-understood system.

Lazebnik starts with the premise that biologists, given their generally feeble knowledge of physics, will look at a radio essentially as a black box that is supposed to play music. Now to figure out how to troubleshoot this complex box, they would start out by securing grants to buy a large supply of identical *functioning* radios. After cataloging the various components ("square metal objects, a family of round brightly colored objects with two legs, round-shaped objects with three legs and so on"), they may shoot the radios at close range and select those for further analysis that develop a malfunction. This may lead to such crucial discoveries as the Most Important Component (the connection between the external FM antenna and the radio) or the Really Important Component (the connection between the internal AM antenna and the radio) or, to the chagrin of the previous two discoverers, the Undoubtedly Most Important Component (the AM/FM switch). Lazebnik ponders:

> *"What is the probability that this radio will be fixed by our biologists? I might be overly pessimistic, but a textbook example of the monkey that can, in principle, type a Burns poem comes to mind."*

(To be fair, I should mention that, in one of his hugely entertaining memoirs, the late physicist Richard Feynman described his failed attempt one year to master the experimental methodology of biologists. As I recall it, Feynman wrote that his biologist colleagues just had a different way of doing science, an approach that he couldn't stay motivated enough to pursue.)

While conceding that the limitations of a purely experimental approach might be somewhat exaggerated in his radio metaphor, Lazebnik contends that the incorporation of a formal language conducive to quantitative analysis (think about electrical schematics and circuit analysis) into the world of experimental biology is the way to go in the future. Lest one should think that Lazebnik's is a lone voice in the wilderness, it is worth noting that the newly emerging field of Systems Biology is perceived to be a paradigm shift by many noted biologists. According to the website [4] of the Institute for Systems Biology:

"Systems biology emerged out of three forces – the Internet, which allows for the transmission, analysis and modeling of large amounts of data; the Human Genome Project and cross-disciplinary science. Systems biology analyzes all of the elements in a system rather than one gene or protein at a time. The systems approach requires the integration of biology, medicine, computation, and technology."

Late CalTech Biologist Eric Davidson noted [1]: "Biology undergoes these revolutionary waves from time to time, after which nothing is ever the same. This is one of those times." Stay tuned!

REFERENCES

[1] S. Begley, "Biologists Hail Dawn of a New Approach: Don't Shoot the Radio," *The Wall Street Journal*, February 21, 2003.

[2] N. Gibbs, "The Secret of Life," *Time*, February 17, 2003.

[3] Y. Lazebnik, "Can a biologist fix a radio—or, what I learned while studying apoptosis," *Cancer Cell*, vol. 2, pp. 179–182, September 2002.

[4] Institute for Systems Biology website [Online]. Available: www.systemsbiology.org (accessed September 1, 2015).

(The original version of the column appeared in "Microwave surfing," *IEEE Microwave Magazine*, vol. 4, no. 4, pp. 28–30, December 2003.)

NOTE

1. To learn how a simple radio really works, see for example: http://electronics. howstuffworks.com/radio8.htm

6.2 THE GRAND CHALLENGES

Notwithstanding Niels Bohr's warning that "prediction is very difficult, especially if it's about the future," I am often drawn (a guilty pleasure?) to our professional colleagues' attempts to gaze into the crystal ball about the future of technologies.

Because of my long-standing interest in biomedical applications, I was especially curious about what the distinguished Fellows of the American Institute for Medical and Biological Engineering (AIMBE) had to say about the future of those fields. AIMBE (www.aimbe.org) celebrated its 20th anniversary in 2011 by undertaking to identify the major trends in the next 20 years. The results of the AIMBE Fellows' discussions were summarized in a recent report [1], which identified the following six overarching challenges:

1. engineering safe and sustainable water and food supply,
2. engineering personalized health care,
3. engineering solutions to injury and chronic diseases,
4. engineering global health through infectious disease prevention and therapy,
5. engineering sustainable bioenergy production,
6. engineering the twenty-first century US economy.

While AIMBE arrived at these recommendations/predictions on its own, somewhat similar conclusions were reached by an IEEE group at a conference held in October 2012. The IEEE group came up with the following five *Grand challenges in Engineering Life Sciences and Medicine* [2]:

1. engineering the brain and nervous system,
2. engineering the cardiovascular system,
3. engineering the diagnostics, therapeutics, and preventions of cancer,
4. translation from bench to bedside,
5. education and training in biomedical engineering.

I drilled a little bit deeper into the AIMBE report [1] to identify areas where engineers working in electromagnetics/RF/microwaves will find technical problems of interest to them within the framework of these grand challenges. Here are a few trends I picked up:

- **Safe and sustainable water and food supply**: Improve the engineering of measurement and control systems (via advances in *sensors, instrumentation*, computing power, and fuzzy/soft system *analytic tools*) for plant, animal, and human quality of life
- **Personalized health care**: Improve medical care by developing expanded capabilities in *telemedicine* and improve early diagnosis and treatment of disease through improved methods for *non-invasive medical imaging*
- **Solutions to injury and chronic diseases**: Improve the treatment of cardiac rhythm disorders by creating *non-invasive pacemakers* (cf. [3])

We may not agree upon all the grand challenges listed by AIMBE or IEEE and we may certainly have our own sense of their relative importance, but it is evident that there will be no shortage of socially relevant technical problems to solve for engineers in our field.

REFERENCES

[1] College of Fellows, AIMBE, "Medical and biological engineering in the next 20 years: the promise and the challenges," *IEEE Transactions on Biomedical Engineering*, vol. 60, no. 7, pp. 1767–1775, July 2013.

[2] B. He, R. Baird, R. Butera, A. Datta, S. George, B. Hecht, et al. "Grand challenges in interfacing engineering with life sciences and medicine," *IEEE Transactions on Biomedical Engineering*, vol. 60, no. 3, pp. 589–598, March 2013.

[3] R. Bansal, "Microwave surfing: tugging at the heart strings," *IEEE Microwave Magazine*, vol. 13, no. 6, pp. 22–141, September/October 2012.

(The original version of the column appeared in "AP-S turnstile," *IEEE Antennas and Propagation Magazine*, vol. 55, no. 4, pp. 218–219, August 2013.)

NOTE

1. Reference [3] is included in this book as Section 6.4.

6.3 BIOMEDICAL APPLICATIONS: TAKING STOCK

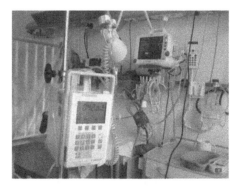

If you attended an *IEEE MTT-S International Microwave Symposium* or an *IEEE International Symposium on Antennas and Propagation and USNC/URSI National Radio Science Meeting* in recent years, you probably noticed that the RF/microwave research community remains seriously engaged in the pursuit of biomedical applications of this part of the electromagnetic spectrum. Such annual events, highlighting the *beneficial* aspects of the non-ionizing electromagnetic waves, clearly remind us of how far we have come in exploring applications that transcend the traditional defense-based boundaries of this engineering field. At the same time, they also serve as a useful counterpoint to the periodic media reports on the putative dangers of non-ionizing electromagnetic radiation.

The IEEE Microwave Symposium in Baltimore [1], which I had the opportunity to attend, showcased biomedical applications of microwaves through technical sessions and workshops. The technical sessions presented new advances in monitoring (e.g., non-contact detection of the vital signs), imaging (including RF transceiver design for magnetic resonance imaging (MRI)), medical sensors (such as those for measuring intracranial pressure), and measurement techniques for the electromagnetic properties of biological tissues. The area of MRI is becoming so important that the symposium had a focused session on high (magnetic) field MRI systems as well as a workshop on MRI systems. A workshop on imaging at millimeter and sub-millimeter wavelengths included a presentation on clinical systems while several workshops explored RF biomedical electronics and sensors (both wearable and implantable) and microwave imaging (e.g., for breast cancer detection).

Similarly, the joint annual IEEE Antennas and Propagation Society symposium and the USNC/URSI National Radio Science meeting in Spokane [2] featured multiple technical sessions devoted to biomedical issues. As would be natural for the sponsoring societies, there was a special emphasis on the role of antennas in biomedical sensor and imaging systems. There were dedicated sessions on biomedical telemetry (including implanted devices), breast cancer detection (with a heavy emphasis on computational modeling), electromagnetic dosimetry and biological exposure assessment, human body interactions with antennas and other electromagnetic devices (including body-centric networks), and therapeutic/rehabilitative applications (e.g., micromagnetic neurostimulation and transcranial magnetic stimulation). In addition, a short course focused on the design of biocompatible antennas for handheld devices (mobile phones, tablet PCs).

Since the research community is often ahead of the curve by virtue of its focus on devices that may be years away from commercial realization, it is also valuable to get a market perspective on medical applications of RF/microwave devices. A report [3] in *Microwaves & RF* is helpful in that regard. While many companies have been in the MRI (the RF transceiver portion) business for years, "an increasing number of firms are enabling and supporting applications like imaging, testing, scanning, and rehabilitation" [3]. One major trend is the gradual replacement of traditionally wired links in the medical systems by wireless ones, generally in the industrial, scientific, and medical (ISM) bands. These links are characterized by low power (running off batteries) and low data rates. IEEE 802.15.4 and Bluetooth Low Energy (BLE) are two of the standards being implemented for wireless medical devices. Beyond the growth in the wireless networking of medical devices, there are current products as well as future opportunities in RF energy sources (e.g., power amplifiers for RF ablation), electric-field sensors for the non-contact reading of electrocardiographs (ECG), and impulse radar transceiver chips (for heartbeat monitoring). For medical applications, the need to get the medical device approved by the Food and Drug Administration (FDA) in the US or by a similar regulatory agency in other countries adds to the complexity of the RF/microwave biomedical device market. While many RF/microwave component/subsystem firms lack direct experience in this area, they are able to play a supporting role for their customers (the medical device/system manufacturers). The report concludes that "despite a steep learning curve posed by FDA approvals and unique application environments, a growing number of RF and microwave firms are finding success serving medical markets" [3].

REFERENCES

[1] 2011 IEEE/MTT-S International Microwave Symposium [Online]. Available: http://www.ieee.org/conferences_events/conferences/conferencedetails/index.html?Conf_ID=14644 (accessed September 2, 2015).

[2] 2011 IEEE International symposium on Antennas and Propagation (APSURSI) [Online]. Available: http://ieeexplore.ieee.org/xpl/mostRecentIssue.jsp?punumber=5981577 (accessed September 2, 2015).

[3] N. Friedrich, "Microwaves Energize Medical Applications," *Microwaves & RF* [Online]. Available: http://mwrf.com/medical/microwaves-energize-medical-applications (accessed September 2, 2015).

(The original version of the column appeared in "AP-S turnstile," *IEEE Antennas and Propagation Magazine*, vol. 53, no. 4, pp. 146–147, August 2011.)

NOTE

1. To start exploring biomedical applications of microwaves, a good starting point is the list of resources listed on the website of IEEE MTT-10, the technical coordinating committee on Biological Effects and Medical Applications: Available: http://www.uta.edu/faculty/jcchiao/MTT_10/Reports.htm (accessed May 4, 2016).

6.4 TUGGING AT THE HEARTSTRINGS

Each year, there are some 80,000 American patients with heart failure [1], who need either a heart transplant or a ventricular assist device (VAD). Since fewer than 2500 heart transplants [2] are performed every year because of a shortage of donors and high costs associated with transplant surgeries, there is a pressing need for improved

VADs. When the first-generation implantable VADs were approved [1] by the FDA in 1994, the expectation was that they would keep the patient alive for a few months while a suitable donor was found. With better VAD technologies, patients with these heart pumps can now survive for 5 or more years [1]. Former Vice President Dick Cheney is one of these VAD recipients.

One nagging problem with VADs persists. Just as a smart phone without a charged battery is little better than an expensive paperweight, implanted VADs need battery power to run them. Because of the power requirements (around 10 W), the power is not supplied by an implanted battery but rather by an external power source through a power line exiting through the chest. This has many adverse consequences: the patient mobility is restricted, it is awkward for the patient to take a shower, and, more seriously, the exit location becomes a potential site of infections, often leading to hospital readmissions for treatment [1, 2].

As I have written before [3–5] about wireless charging of consumer electronics, inductive coupling is common for household devices such as electric toothbrushes but the power transfer efficiency of this approach drops off rapidly as the distance between the transmitter and the receiver coils increases. A version of this strategy, called TETS (transcutaneous energy transfer systems) was implemented for VADs with power transfer coils located below and above the skin. After extensive laboratory testing and prototype development, two TETS were commercially implemented and marketed for clinical use. Unfortunately, these schemes have low tolerance for misalignment between the two coils and the placement of the external coil on the skin can result in local irritation, infection, and even thermal injury [1].

Against the above background, a recent collaboration between a heart surgeon and an electrical engineer [1, 6] generated a lot of excitement in the media as well as in professional forums. Their system called Free-Range Resonant Electrical Energy Delivery System (FREE-D) is also based on induction but uses resonant coupling between transmitter and receiver coils tuned to a specific frequency in the MHz range, allowing a power transfer with high efficiency over a distance of the order of a meter (rather than a few millimeters as in typical inductive coupling setups). The concept of resonant non-radiative coupling has been explored before but the team of Bonde and Smith [1, 6] was the first one to exploit it for powering VADs wirelessly. In their proposed implementation, one small coil (receiver) can be implanted in the body while the larger transmitting coil can be worn conveniently in a vest or even attached to a bed (at night) or mounted on the ceiling. While the laboratory tests of the FREE-D system attest to its great potential, actual clinical use must await animal studies and clinical trials. If all goes well, in a few years patients with heart failure can look forward to having a wireless lifeline.

REFERENCES

[1] B. H. Waters, A.P. Sample, P. Bonde, and J. R. Smith, "Powering a ventricular assist device (VAD) with the free-range resonant energy delivery (FREE-D) system," *Proceedings of the IEEE*, vol. 100, no. 1, pp. 138-149, January 2012.

[2] "A wireless heart," *The Economist* [Online]. Available: http://www.economist.com/node/21017837 (accessed September 2, 2015).

[3] R. Bansal, "AP-S turnstile: cutting the cord," *IEEE Antennas and Propagation Magazine*, vol. 49, no. 1, p. 150, February 2007.

[4] R. Bansal, "Microwave surfing: goodbye to batteries," *IEEE Microwave Magazine*, vol. 8, no. 4, pp. 24–26, August 2007.

[5] R. Bansal, "Microwave surfing: the future of wireless charging," *IEEE Microwave Magazine*, vol. 10, no. 5, p 30, August 2009.

[6] The website for Prof. J. Smith's Sensor Systems Laboratory at the University of Washington [Online]. Available: http://sensor.cs.washington.edu/FREED.html (accessed September 2, 2015).

(The original version of the column appeared in "AP-S turnstile," *IEEE Antennas and Propagation Magazine*, vol. 54, no. 3, pp. 170–171, June 2012.)

NOTE

1. Reference [5] is included in this book as Sections 8.7.

6.5 A JOLT FROM THE BLUE?

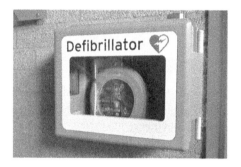

"Researchers demonstrate ability to reprogram wireless signals from defibrillators, steal patient information" [1].

—Headline of a CNN story

Pacemakers and implantable cardioverter defibrillators (ICDs) are routinely implanted in US patients [2], saving and prolonging many lives by helping regulate cardiac function. When one of these devices (newer models) is surgically placed below the patient's skin, it has the capability of communicating with a nearby external programming unit both for telemetering data from the patient (physiological data as well as identification information) and for receiving instructions that modify the operational parameters of the device, frequently over an *unencrypted* channel. There's the rub.

A team of researchers led by Dr. William Maisel of Harvard Medical School and including electrical engineers and computer scientists from University of Massachusetts (Amherst) and the University of Washington decided to "hack" the wireless system to demonstrate the potential vulnerabilities of the implanted medical devices. In a widely publicized paper [2] presented at the IEEE Symposium on Security and Privacy (May 2008), the team reported its investigation of a wireless-enabled (175 kHz) Medtronic defibrillator unit. The team started by reverse-engineering the communication protocol used by the ICD with the help of a fast (4 GSa/s) recording oscilloscope and a Universal Software Radio Peripheral (USRP). The oscilloscope recorded RF traces both from the ICD and the nearby programming unit. These traces were processed in a Matlab environment with the help of the GNU radio toolchain to recover the bits being transmitted back and forth between the ICD and the external programming unit.

Armed with the communication protocol, the team proceeded not only to eavesdrop successfully on the telemetered data from the ICD (which would compromise the patient's records) but also to launch software radio-based attacks on the ICD (which could affect the patient's safety) by (for example) administering unneeded shocks to the heart. Before you get too upset by the team's "callous" approach, let me add that no patients were used in the study (only a slab of bacon and some ground beef in some of the experimental work) and the paper [2] omits many technical details to prevent unscrupulous readers from trying the hacking in a clinical setting. Since the team's primary motivation was to improve patient safety, the researchers also described their approaches to zero-power (no battery power needed) defenses that would alert the patient in case of a software radio attack (e.g., by providing an audible signal that draws its power from RF power harvesting). The zero-power concept is essential in any defense strategy since power consumption is a serious issue for a battery powered implanted device.

Not everyone was impressed. Bruce Lindsay of Cleveland Clinic noted [1]: "[Defibrillator transmissions are] not designed to withstand terrorist attacks....I don't think the findings have any great clinical significance. To hack the system, you have to get the programmer right up against the patient's chest. It is not as if somebody could do this from down the street." To which Dr. Maisel's (leader of the study) response was: "There will be more implanted devices and more wireless capabilities and transmissions over greater distances." A Medtronic spokesman agreed that future versions of ICDs would, indeed, transmit signals as far away as 30 feet from the patient but they *would* incorporate stronger security. If that does happen, the study would have served its purpose.

REFERENCES

[1] "Study: Heart Devices can be Hacked," was a 2008 news story reported on CNN.

[2] D. Halperin, T. S. Heydt-Benjamin, B. Ransford, S. S. Clark, B. Defend, W. Morgan, et al., "Pacemakers and implantable cardiac defibrillators: software radio attacks and zero-power defenses," *Proceedings of the 2008 IEEE Symposium on Security and Privacy*, pp. 129–142, May 2008.

(The original version of the column appeared in "AP-S turnstile," *IEEE Antennas and Propagation Magazine*, vol. 50, no. 6, p. 152, December 2008.)

NOTE

1. To learn more about software defined radio, see for example: http://www.arrl.org/software-defined-radio (accessed May 4, 2016).

6.6 CHANNELING THE VOICE WITHIN

A century after Marconi's pioneering experiments in radio communication and about 50 years after the birth of the microprocessor, the marriage of the two technologies has led to over four billion mobile phone users around the world. Since Marconi's days, the radio spectrum efficiency has increased perhaps a *trillion* times (as estimated by the wireless pioneer Martin Cooper) just as the latest generation of microprocessors attests to the relentless march of Moore's law with over a *billion* transistors per chip, and, as a result, the estimated cost of a wireless system has plummeted to a *penny* per delivery [1].

If every shopper walking in a mall is already talking on a mobile phone and if every gadget you buy already comes equipped with an embedded wireless chip, what is the next frontier in wireless connectivity? Wireless chips are finding their way *into* people's bodies. For example, patrons wishing to enter the VIP area of the Baja Beach Club in Barcelona had identification in the form of microchips implanted in their arms [1]. On the medical side, as Mark Norris predicted in *Design News* [2], wireless implants, including cardiac pacemakers, cochlear implants, and neurostimulators, represented a $17 billion market.

In 1999, the Federal Communications Commission (FCC) [3] established Medical Implant Communications Service (MICS) in the 402–405 MHz frequency band as "an ultra-low power, unlicensed, mobile radio service for transmitting data in support of diagnostic or therapeutic functions associated with implanted medical devices."

The 402–405 MHz frequency band was chosen because of the following attractive features:

- Good radio signal propagation characteristics within the human body
- Suitability for meeting the MICS requirements in terms of size, power, antenna performance, and receiver design
- Compatibility with international frequency allocations

In order to avoid interference with other users in the band, MICS devices have to meet stringent specifications [3]:

- Effective isotropic radiated power (EIRP) no greater than 25 microwatts
- Limited to an authorized bandwidth of 300 kHz
- May not be used for voice communications
- Must be tested for emissions and EIRP limit compliance while enclosed in a medium that simulates human body tissue
- Frequency stability of ±100 ppm of the operating frequency

As Mark Norris [2] noted, meeting the FCC requirements while providing a highly reliable data communication service without consuming significant battery power is a "delicate balancing act." It forces researchers and manufacturers to optimize all aspects of the systems from the antenna [4] to the transceiver chip [5]. For example, the popular Zigbee standard can setup and manage *ad hoc* mesh networks and is suitable for *external* medical devices in a hospital environment but it does not meet the ultra-low power consumption requirement of implantable devices [2]. The other challenge is economic. As Maarten Barmentlo, the chief technology officer of Philips's consumer health-care division emphasized [1], "the basic technology to make these things happen exists; the big issue is how to make this economically viable." But these are the early days for "Wireless In*corp*orated," the future of the bionic man is definitely wireless. Leave your radio dial tuned to 402–405 MHz.

REFERENCES

[1] "A World of Connections," *The Economist* [Online]. Available: http://www.economist.com/node/9032088 (accessed September 2, 2015).

[2] M. Norris, "Design considerations for Wireless Implants," *Design News* [Online]. Available: http://www.designnews.com/document.asp?doc_id=220968 (accessed September 2, 2015).

[3] Background about wireless medical devices at the FCC website [Online]. Available: https://www.fcc.gov/page/fcc-connect-2-health-mobile-healthcare-technology-milestones (accessed September 2, 2015).

[4] A. Mahanfar, S. Bila, M. Aubourg, and S. Verdeyme, "Design Considerations for the Implanted Antennas," *IEEE MTT-S International Microwave Symposium*, pp. 1353–1356, Honolulu, HI, June 3–8 2007.

[5] R. Merritt, "Zarlink Transceiver Helps Implants go Broadband," *EETimes* [Online]. Available: http://www.eetimes.com/news/latest/showArticle.jhtml?articleID=199203169 (accessed September 2, 2015).

(The original version of the column appeared in "AP-S turnstile," *IEEE Antennas and Propagation Magazine*, vol. 49, no. 3, pp. 160–161, June 2007.)

NOTE

1. For a technical discussion of the field of wearable and implanted RF sensors and networks, see for example:
D. H. Werner and Z. H. Jiang, *Electromagnetics of Body-Area Networks: Antennas, Propagation, and RF Circuits*, Wiley-IEEE Press, 2016.

6.7 BATTLING CANCER: MICROWAVE HYPERTHERMIA

Hyperthermia is a procedure for treating cancer in which the tissue temperatures in the tumor are raised to the range of 42–45°C [1]. *Microwave* hyperthermia was introduced several decades ago (see, e.g., Paglione's US patent [2] and the references cited therein). It has been used generally in combination with chemotherapy and/or radiotherapy because hyperthermia enhances the tumor's response to radio/chemotherapy. An example of experimental research on microwave hyperthermia is the work by Maccarini et al. [3], presented in the technical session "Biological Effects and Medical Applications" at IMS 2005. However, as Maccarini et al. [3] noted, "Microwave hyperthermia...survived mostly as academic research, because of the difficulties involved with commercially available heating equipment and clinicians' skepticism." This treatment modality has seen more clinical usage in Europe and Asia [4, 5]. In the United States, microwave hyperthermia in conjunction with radiation has received approval from the FDA, but [4] "few doctors are trained to use it." The publication [6, 7] of the results of a clinical trial may help change the US scene.

The aforementioned US clinical study was carried out at Duke University by a team led by Dr. Ellen Jones. The team felt that earlier hyperthermia studies had generally lacked "rigorous thermal dose prescription and administration." In its clinical trial, the Duke team used "the number of cumulative equivalent minutes exceeded by 90% of monitored points within the tumor (CEM 43°C T_{90}) as a measure of thermal dose." The trial involved 109 patients with superficial ($\leq$3 cm depth) "heatable" tumors. Most of the patients were women with breast cancer that had recurred in the chest wall following surgery.

Based on prior preclinical as well as clinical data, the investigators' estimated that the minimum effective thermal dose was 10 CEM 43°C T_{90}. The hyperthermia plan involved twice-a-week sessions of 1–2 hour duration for a maximum of 10 treatments. It was delivered using externally located microwave spiral strip applicators operating at 433 MHz. Fiber optic thermometers were used to monitor the temperature distribution in the tumor. The hyperthermia was combined with radiotherapy. It was a randomized trial, with some patients receiving only radiotherapy while others received the combination of hyperthermia and radiotherapy.

The dual therapy was found to be completely effective in destroying the tumor in 66% of the patients receiving radiotherapy and hyperthermia while the success rate was only 42% for those receiving radiotherapy by itself. Patients who had received radiotherapy previously had the greatest incremental gain in complete response: 68% for the dual-therapy cases versus only 24% for those receiving no hyperthermia. Overall, the dual therapy did *not* extend survival for the patients, "primarily because so many patients had metastases elsewhere in the body." However, by controlling the tumor locally, the combined therapy was critical to enhancing the patient's quality of life. It also provided a strategy for getting "more mileage out of a modest dose of radiation for previously treated patients, who cannot tolerate a full dose" [7].

REFERENCES

[1] J. C. Lin, "Biomedical Applications of Electromagnetic Engineering," Chapter 17, pp. 605–629, In *Handbook of Engineering Electromagnetics*, edited by R. Bansal, Marcel Dekker: New York, 2004.

[2] R. Paglione, "Coaxial Applicator for Microwave Hyperthermia," US Patent #4,204,549, May 27, 1980.

[3] P. F. Maccarini, H.-O. Rolfsnes, D. G. Neumann, J. Johnson, T. Juang, and P. R. Stauffer, "Advances in Microwave Hyperthermia of Large Superficial Tumors," *IEEE MTT-S International Microwave Symposium*, Long Beach, CA, June 12–17, 2005.

[4] N. Seppa, "Microwavable cancers," *Science News*, vol. 167, no. 19, p. 294, May 7, 2005.

[5] A. G. van der Heijden, L. A. Kiemeney, O. N. Gofrit, O. Nativ, A. Sidi, Z. Leib, et al., "Preliminary European results of local microwave hyperthermia and chemotherapy treatment in intermediate or high risk superficial transitional cell carcinoma of the bladder," *European Urology*, vol. 46, no. 1, pp. 65–71, July 2004.

[6] E. L. Jones, J. R. Oleson, L. R. Prosnitz, T. V. Samulski, Z. Vujaskovic, D. Yu, et al., "Randomized trial of hyperthermia and radiation for superficial tumors," *Journal of Clinical Oncology*, vol. 23, no. 13, pp. 3079–3085, May 1, 2005.

[7] A Duke University press release on the study [6].
(The original version of the column appeared in "Microwave surfing," *IEEE Microwave Magazine*, vol. 6, no. 3, pp. 32–34, September 2005.)

NOTE

1. For an overview of the use of hyperthermia (including microwave hyperthermia) in the treatment of cancer, see for example: http://www.cancer.org/treatment/treatmentsandsideeffects/treatmenttypes/hyperthermia (accessed May 4, 2016).

DID YOU KNOW?

A FUN QUIZ (VI)

1. For the 1988 IEEE/MTT-S Hertz Centennial Celebration, IEEE published a commemorative volume *Heinrich Hertz: The Beginning of Microwaves*. Recently, while preparing the introductory lecture for my graduate course on antennas, I reread the volume and once again marveled at Hertz's experimental ingenuity. The first two questions are derived from that volume. The phenomenon of *skin effect*, that is, the tendency of high-frequency currents to flow in a narrow layer of a conducting body, was explicitly suggested by:
   (a) Maxwell
   (b) Faraday
   (c) Hertz
   (d) None of the above

2. Starting in 1886, Hertz (1857–1894) conducted successful experiments to validate Maxwell's theory (1864) while Hertz was:
   (a) A doctoral student under Helmholtz at the University of Berlin
   (b) An instructor at the University of Kiel
   (c) A Professor at Karlsruhe
   (d) None of the above

3. *mHealth* is:
   (a) The field dealing with the potential health hazards of mobile phones
   (b) The use of magnetic devices, for example, bracelets, to achieve putative health benefits
   (c) The use of new mobile phone technology applications to provide health services and information
   (d) None of the above

4. *Disestimation* involves:
   (a) Assigning too much meaning to a measurement, relative to the uncertainties and errors inherent in it
   (b) Removing the part of a radar return associated with clutter
   (c) Expunging contradictory signals in data fusion
   (d) None of the above

From ER to E.T.: How Electromagnetic Technologies Are Changing Our Lives, First Edition. Rajeev Bansal.
© 2017 by The Institute of Electrical and Electronic Engineers, Inc. Published 2017 by John Wiley & Sons, Inc.

5. A research group at the National Institute of Standards (NIST) reported measuring the smallest force ever: 174 yoctonewtons, which corresponds to:
   (a) 174×10^{-18} N
   (b) 174×10^{-21} N
   (c) 174×10^{-24} N
   (d) None of the above

6. According to the British tabloid *Sun*, Lady Gaga is afraid to use:
   (a) A cell phone
   (b) An iPad
   (c) The internet
   (d) None of the above

7. In a magazine column in 1995, network engineer Bob Metcalfe predicted that the exponential growth in data traffic would cause the internet to collapse within a year. When the apocalypse did not arrive, Metcalfe:
   (a) Wrote another column predicting the collapse of the internet within a decade
   (b) Literally ate his words by mixing a copy of the column with water in a blender and swallowing it
   (c) Issued an apology through his Twitter account
   (d) None of the above

8. A German research group has harnessed the energy of electromagnetic pulses (EMP) for:
   (a) Disabling public Wi-Fi hot spots
   (b) Reducing pollutants in the air inside a car
   (c) Punching holes through steel
   (d) None of the above

9. A research group from the North Carolina State University has proposed networking distributed sensors in buildings via:
   (a) Radio waves through heating, ventilation, and air conditioning (HVAC) ducts
   (b) Infrared signals reflected from the windows
   (c) Overlaying the signals on the 60-Hz wiring
   (d) None of the above

10. A team from Stanford University has developed an "*e-shirt*," which:
    (a) Displays an electronically alterable message
    (b) Uses carbon nanotube-based inks to dye the shirt and turn it into a battery
    (c) Is a virtual t-shirt based on optical metamaterials
    (d) None of the above

ANSWERS

1. (a) Maxwell
   Source: J. Bryant, *Heinrich Hertz: The Beginning of Microwaves*, IEEE, NY, 1988.

2. (c) A Professor at Karlsruhe
   Source: J. Bryant, *Heinrich Hertz: The Beginning of Microwaves*, IEEE, 1988.

3. (c) The use of new mobile phone technology applications to provide health services and information
   Source: K. Foster, "Telehealth in Sub-Saharan Africa: lessons for humanitarian engineering," *IEEE Technology and Society Magazine*, pp. 42–49, Spring 2010.

4. (a) Assigning too much meaning to a measurement, relative to the uncertainties and errors inherent in it
   Source: C. Seife, *Proofiness: The Dark Arts of Mathematical Deception*, Viking, 2010.

5. (c) 174×10^{-24} N
   Source: "The Force is Weak With This One," *The Economist*, p. 78, April 24, 2010.

6. (a) A cell phone
   Source: "'*Scientific American*' v. *Lady Gaga*," *Microwave News* [Online]. Available: http://www.microwavenews.com/news-center/%E2%80%9Cscientific-american%E2%80%9D-vs-lady-gaga (accessed September 3, 2015).

7. (b) Literally ate his words by mixing a copy of the column with water in a blender and swallowing it
   Source: "Breaking Up," *The Economist*, p. 65, February 13, 2010.

8. (c) Punching holes through steel
   Source: "It's a Knockout," *The Economist*, p. 80, January 16, 2010.

9. (a) Radio waves through heating, ventilation, and air conditioning (HVAC) ducts
   Source: C. Dillow.. "Sensor Networks in Buildings Could Use AC Ducts as Huge Building-Wide Antennas," *Popular Science* [Online]. Available: http://www.popsci.com/science/article/2010-08/rfid-sensor-networks-buildings-would-use-ac-ducts-huge-building-wide-antennas (accessed September 3, 2015).

10. (b) Uses carbon nanotube-based inks to dye the shirt and turn it into a battery
    Source: Dye turns fabric into a battery. *BBC News* [Online]. Available: http://news.bbc.co.uk/2/hi/technology/8471362.stm (accessed September 3, 2015).

(The original version of the quiz appeared in "AP-S turnstile," *IEEE Antennas and Propagation Magazine*, vol. 53, no. 1, pp. 129–130, February 2011.)

CHAPTER 7

DEFENSE APPLICATIONS

"We've opened the door into the secret garden."

—Sir John Pendry (1943–)

7.1 WHERE IS WALDO?

"We've opened the door into the secret garden."

Professor Pendry's (Imperial College, London) observation [1] from 2006 is an example of the excitement in the electromagnetic research community on the potential of

From ER to E.T.: How Electromagnetic Technologies Are Changing Our Lives, First Edition. Rajeev Bansal.
© 2017 by The Institute of Electrical and Electronic Engineers, Inc. Published 2017 by John Wiley & Sons, Inc.

metamaterials to create the equivalent of Harry Potter's invisibility cloak. Back then, a research group at Duke University demonstrated [2] how a 2-dimensional structure composed of 10 fiberglass rings covered with sub-wavelength arrays of copper elements could help "channel" an incident EM wave in the microwave frequency range around a copper cylinder, minimizing the scattered signal. In 2008, a group at UC Berkeley [3] raised the stakes by fabricating the "cloak" using nanotechnology (for the sub-wavelength elements) so that the target could be made to "disappear" at wavelengths approaching the visible spectrum. Exciting as these metamaterial-based developments have been, they remain far from real-life applications. In the meantime, a report in *The Economist* [4] discusses what is achievable in practice using other hide-and-seek military technologies. Here are some highlights from the report [4]:

Battledress uniforms ("fatigues") have traditionally been printed with wiggly patterns of solid colors ("tiger stripes"). Researchers in the field of "clutter metrics" (the study of observers' ability to locate and identify targets) have used eye-movement tracking to discredit the tiger stripes in favor of fabrics with small squares of colors ("pixels"). These pixel patterns, already in use on many western army uniforms, force observers to be 40% closer in order to spot the camouflaged soldiers.

Camouflage that dynamically adapts to the environment is another technology being explored. One version incorporates a small camera to scan the surroundings. The resulting data is used to manipulate the colors and patterns of textile-like plastic sheets embedded with light-emitting diodes (LEDs). While the sheets are not yet "wearable," they are being tested for camouflaging equipment. For example, a tank cloaked in such sheeting and parked in front of a grassy slope, displays a grassy image on the front. In another version of adaptive camouflage, the color-and-pattern feedback from the camera is used to project crude replicas on flexible plastic decals (rudimentary computer displays) affixed to equipment.

For nighttime operations, it becomes more important to conceal the heat signature of the equipment from infrared and thermal-vision equipment. Fabrics with embedded cenospheres (tiny hollow spheres of aluminum and silica) are being tested for their effectiveness in blocking body-heat signatures. Insulating decals that can be applied to hot objects (e.g., artillery cannons) are also in the works. While they may not conceal the target completely, they can confuse the observer by making a tank appear like a motorbike in terms of its heat signature.

On the detection side in addition to ongoing work on foliage-penetrating radar, an interesting development takes advantage of the radio "chatter" background created by cell phones in an urban environment. When an aircraft (even with "stealth" characteristics) flies through this "chatter," it creates a "hole" in the background, which can be detected. What makes the technique even more appealing is that it is a "passive" detection approach, leaving the detection system relatively unexposed. Thus continues the hi-tech game of hide-and-seek.

REFERENCES

[1] P. Rincon, Experts create invisibility cloak, *BBC News* [Online]. Available: http://news.bbc.co.uk/2/hi/science/nature/6064620.stm (accessed September 3, 2015).

[2] D. Felbacq, "Envisioning invisibility: recent advances in cloaking," *OPN*, vol. 18, no. 6, pp. 32–37, June 2007.

[3] "Invisibility Cloak 'Step Closer'," *BBC News* [Online]. Available: http://news.bbc.co.uk/2/hi/science/nature/7553061.stm (accessed September 3, 2015).

[4] "How to Disappear," *The Economist* [Online]. Available: http://www.economist.com/science/tq/displaystory.cfm?story_id=11999355 (accessed September 3, 2015).

(The original version of the column appeared in "AP-S turnstile," *IEEE Antennas and Propagation Magazine*, vol. 50, no. 5, p. 156, October 2008.)

NOTE

1. To learn more about metamaterials, see for example: http://phys.org/tags/metamaterials/ (accessed May 5, 2016).

7.2 ANTIMAGNET

"It is quite amazing that almost 160 years after Maxwell equations were first developed, we are still finding new solutions based solely on them!" [1].

—Alvaro Sanchez
Universitat Autònoma de Barcelona, Spain

Can Maxwell equations be harnessed to design an "antimagnet" [1, 2], a cloak that would, on the one hand, conceal the *static* magnetic field produced by an object inside and, on the other, would not perturb an external *static* magnetic field? If such a cloak were to be realized, one could imagine passing a forbidden metal object undetected through a metal detector (thereby creating a security nightmare) or, on a grander scale, hiding a submarine from a magnetically triggered underwater mine.

Inspired by earlier work [3] from John Pendry's metamaterials group at Imperial College (London), Alvaro Sanchez (Universitat Autònoma de Barcelona) and his colleagues have reported in *Science* [4] both the theoretical design for an "antimagnet" and an experimental set up to demonstrate the concept:

> *"Invisibility to electromagnetic fields has become an exciting theoretical possibility. However, the experimental realization of electromagnetic cloaks has only been achieved starting from simplified approaches (for instance, based on ray approximation, canceling only some terms of the scattering fields, or hiding a bulge in a plane instead of an object in free space). Here, we demonstrate, directly from Maxwell equations, that a specially designed cylindrical superconductor-ferromagnetic bilayer can exactly cloak uniform static magnetic fields, and we experimentally confirmed this effect in an actual setup"* [4].

The new "antimagnet" cloak has an inner layer fabricated from a high-temperature superconducting tape, which, on its own, repels the magnetic field. The outer ferromagnetic layer is made from a thick FeNiCr alloy sheet and, on its own, it draws in the magnetic field lines. The dimensions and the material properties are computed based on Maxwell equations so that the counteracting effects of the layers are balanced to eliminate any distortion of the external static magnetic field. Since for a static (dc) situation, the magnetic and electric fields decouple, the design of the cloak involves only the magnetic permeability of the materials.

The Sanchez group tested their "antimagnet" cloak on a small scale (12.5×12 mm) in a static magnetic field of 40 mT [1]. Since the design is magnetostatic, there is no wavelength dependence and, in principle, the cloak can be scaled up and down in size. However, the so-called high-temperature superconducting tape used in the fabrication requires liquid nitrogen temperatures, so real-life applications to metal detectors and submarines are still futuristic.

REFERENCES

[1] T. Commissariat, "How to Hide From a Magnetic Field," *Physicsworld* [Online]. Available: http://physicsworld.com/cws/article/news/2012/mar/22/how-to-hide-from-a-magnetic-field (accessed October 22, 2015).

[2] J. Palmer, "'Antimagnet' Joins List of Invisibility Approaches," *BBC News* [Online]. Available: http://www.bbc.co.uk/news/science-environment-15017479 (accessed October 22, 2015).

[3] F. Magnus, B. Wood, J. Moore, K. Morrison, G. Perkins, J. Fyson, et al., "A d.c. magnetic metamaterial," *Nature Materials*, vol.7, pp. 295–297, 2008.

[4] F. Gömöry, M. Solovyov, J. Šouc, C. Navau, J. Prat-Camps, and A. Sanchez, "Experimental realization of a magnetic cloak," *Science*, vol. 335, no. 6075, pp. 1466–1468, March 2012.

(The original version of the column appeared in "AP-S turnstile," *IEEE Antennas and Propagation Magazine*, vol. 54, no. 5, p. 198, October 2012.)

NOTES

1. Metamaterials are discussed also in Section 7.1.
2. Textbook resources:
   (i) W. H. Hayt and J. A. Buck, *Engineering Electromagnetics*, 8th ed., McGraw-Hill, New York, 2012. Magnetic materials are discussed in Chapter 8.
   (ii) F. T. Ulaby and U. Ravaioli, *Fundamentals of Applied Electromagnetics*, 7th ed., Prentice Hall, Upper Saddle River, NJ, 2015. Magnetic materials are discussed in Chapter 5.

7.3 CUTTING TO THE CHASE

In 2004, on the NPR (National Public Radio) program *All things Considered*, Robert Siegel interviewed [1] Dr. David Giri, president of a small California-based consulting company ProTech, about a new microwave device that could spell the end of hair-raising car chases, so beloved of Hollywood film makers. A prototype had been commissioned by law enforcement agencies and police forces in the United States and Britain had ordered tests of the new system [2]. Results of early trials of the device were presented at the EUROEM 2004 conference [3].

As Dr. Giri explained in the NPR interview, the device is designed to fit in the trunk of a police car with a directive antenna mounted on the roof. At the touch of a button, a police officer can direct a jolt of microwave energy at a speeding car. The electromagnetic impulse induces high transient currents in the wires leading to the microprocessor of the suspect's car and disables the ignition. Even though BMWs and Toyotas may use microprocessors from different manufacturers, they are sufficiently similar in terms of their susceptibility to a microwave pulse that the police officer does not have to worry about using different settings. The main technical challenge is to make the unit directional enough so it is only the O. J. Simpson SUV (oops! I mean the suspect's car) that is stopped in its track and not all the other cars sharing the same freeway. I would have thought that if a car traveling at a high speed suddenly loses its electronic ignition, it may have a tendency to veer out of control and cause a massive pileup on the road. But Dr. Giri assured the NPR interviewer that such was not the case and the targeted car would merely continue to slow down and eventually stop. Another concern would be the potential harm the microwave pulse might do to the occupants of the suspected car. After all, the US Department of Defense has demonstrated a microwave system that can be used to subdue a hostile crowd by causing temporary painful heating with directed microwave energy [4]. Once again, Dr. Giri was reassuring about his device, claiming it would comply with the applicable IEEE standard for safe exposure to microwaves.

While the system developed by Dr. Giri has been commissioned by law enforcement agencies, it does not require much of a stretch in imagination that a device like this can easily find its way into the hands of miscreants who can use it to sabotage all kinds of electronic gear controlled by microprocessors. Dr. Giri readily conceded that point and noted that this was, in fact, one of the issues addressed in his new book [5] on non-lethal electromagnetic weapons.

If Dr. Giri's car zapper catches on with police departments, what is a compulsive bank robber like Willie Sutton supposed to do? Let's cut right to the chase: a 1965 Ford Mustang has no microprocessor onboard.

REFERENCES

[1] The NPR interview is available online in the NPR archives through http://www.highbeam. com/doc/1P1-96651030.html (accessed September 15, 2015).

[2] Ian sample, "Police Test Hi-Tech Zapper That Could End Car Chases," *The Guardian* (UK) [Online] Available: http://www.theguardian.com/science/2004/jul/12/sciencenews.crime (accessed September 15, 2015).

[3] EUROEM 2004, July 12–16, 2004, Magdeburg, Germany.

[4] S. Mihm, "The Quest for the Nonkiller App," *The New York Times*, July 27, 2004.

[5] D. V. Giri, *High-Power Electromagnetic Radiators: Nonlethal Weapons and Other Applications*, Harvard University Press, Cambridge, MA, December 2004.

(The original version of the column appeared in "Microwave surfing," *IEEE Microwave Magazine*, vol. 6, no. 1, p. 36, March 2005.)

NOTES

1. The company e2v has developed a product called "RF Safe-Stop" based on this concept. See: http://www.e2v.com/products/rf-power/rf-safe-stop/ (accessed May 5, 2016).

2. Section 7.4 also discusses work by Dr. Giri on non-lethal electromagnetic weapons.

7.4 TWENTY-FIRST CENTURY WARFARE

One of the questions on the IMS 2005 Quiz, which appeared in the June 2005 issue of the *IEEE Microwave Magazine*, related to the applications of high-power electromagnetic technology. In particular, it asked:

Active Denial Technology uses microwaves to…

(a) Prevent unauthorized access to the protected premises

(b) Disperse a hostile crowd

(c) Jam enemy communications

(d) None of the above

The correct answer, (b) disperse a hostile crowd, referred to the work the US Air Force had done to develop a non-lethal microwave weapon (*active denial technology* [1]) to disperse hostile crowds by beaming microwaves at specific individuals. The heat generated by the beam is supposed to create a temporary burning sensation without producing any adverse long-term effects.

Section 7.3 discussed another application of high-power microwave technology. The device, which is designed to fit in the trunk of a police car with a directive antenna mounted on the roof, can direct a jolt of microwave energy at a speeding car. The electromagnetic impulse induces high transient currents in the wires leading to the microprocessor of the suspect's car and disables the ignition.

At an alumni lunch to celebrate late Professor R. W. P. King's (Harvard University) 100th birthday in 2005, I was seated next to D. V. Giri, and had a chance to talk with him about his new book [2] on non-lethal electromagnetic weapons. The book, which represented the first entry in the series "The Electromagnetics Library" by the Harvard University Press, examines various non-lethal weapons (NLW) technologies, emphasizing those based on high-power electromagnetic pulses (EMPs). Giri, a consulting scientist based in California, has extensive experience with EMP simulators, and has co-authored a previous book [3] on high-power microwave systems.

On the one hand, Giri's new book is a brief survey of current and emerging technologies of several types of sources and radiating systems for narrow, moderate, ultra-moderate, and hyper-band (band ratio greater than a decade) high-power electromagnetic signals. On the other hand, it is a polemic advocating the view that the twenty-first century will call for a new paradigm for weapons technology and that NLW are going to play an increasingly important role in combat and civil conflicts in the coming years. As Giri sees it, a credible argument for developing NLW technologies is based on the following considerations [2]:

1. The United States has become the sole superpower without a symmetric adversary.
2. The United States, a major player in NATO, places its personnel in risky situations during peacekeeping operations.
3. Major western countries prefer solutions that minimize the loss of life on both sides of a conflict.
4. Law enforcement agencies are natural beneficiaries of NLW technologies.
5. NLW technologies are a way of controlling dissent and insurgencies without increasing antagonism (as might be the case when conventional weapons are aggressively deployed).
6. NLWs are better suited to urban warfare (Kosovo, Baghdad, etc.).
7. NLW technologies have the potential of decreasing the risks of friendly fire.

Giri, despite his high enthusiasm for NLW technologies, was pragmatic enough to concede that the NLW technologies, even if widely developed and deployed, would not replace conventional weapons but would play a *complementary* role. Also, NLW have their own limitations [2]: skepticism ("the concept of non-lethal is antithetical to war itself"), the potential development cost, possible misuse in the wrong hands, and, above all, the vulnerability of our whole computer and electronic infrastructure to large-scale high-power electromagnetic weapons used by the enemy forces. Of course, that naturally leads us to the topic of non-lethal technology *countermeasures*. The topic of Giri's next book, perhaps?

REFERENCES

[1] S. Mihm, "The Quest for the Nonkiller App," *The New York Times*, July 27, 2004.
[2] D. V. Giri, *High-Power Electromagnetic Radiators: Nonlethal Weapons and Other Applications*, Harvard University Press, Cambridge, MA, December 2004.
[3] C. D. Taylor and D. V. Giri, *High-Power Microwave Systems and Effects*, Taylor and Francis, 1994.

(The original version of the column appeared in "Microwave surfing," *IEEE Microwave Magazine*, vol. 7, no. 1, pp. 36–38, February 2006.)

7.5 ON A WING AND A PRAYER

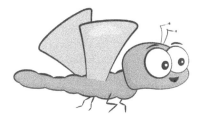

Look below, there's our field over there

Though there's one motor gone

We can still carry on

Comin' in on a wing and a prayer.

These lines from a popular World War II song [1], with words by Harold Adamson and music by Jimmy McHugh, glided into my mind as I read, over a Memorial Day weekend, a DARPA (Defense Research Projects Agency) pre-solicitation notice [2] in the area of Hybrid insect MEMS (microelectromechanical systems).

DARPA's goal was to fund promising interdisciplinary research that will permit the creation of "insect-cyborgs." Proceeding from the premise that the whole is greater than the sum of the constituent parts, DARPA would have liked to have MEMS intimately integrated into insects, during their early stages of metamorphoses. As the pupa develops, the incision needed to insert the MEM will presumably heal and the internal organs will develop around the MEM. This should lead, in principle, to a more reliable bio-electromechanical interface to the insect, as compared with the earlier DARPA-funded attempts [3] to glue the electronic module to an adult insect.

Once the holistic integration of an MEM with the insect was perfected, the focus would shift to "controlling insect locomotion and sense local environment" (e.g., for detecting explosives or landmines). Since finding a light enough power source for the MEM can be difficult, power scavenging (from body heat or movement) was considered another desideratum. The final demonstration goal of the DARPA program was the "delivery of an insect within five meters of a specific target located at hundred meters away, using electronic remote control, and/or global positioning system (GPS)." Although DARPA's primary interest was flying insects (e.g., moths and dragonflies), hopping and swimming insects were also acceptable platforms.

Only time will tell whether this is just another blue-sky scheme or whether it will produce some tangible results for DARPA. Speaking to the BBC, Dr. George McGavin, an entomologist affiliated with the Oxford University Museum of Natural History, expressed his skepticism [4]: "What adult insects want to do is basically reproduce and lay eggs. You would have to rewire the entire brain patterns [to control flight patterns]." Previous DARPA-funded projects on bees and wasps ran into difficulties [4] when the insects' "instinctive behaviors for feeding and mating…prevented them from performing reliably." Of course, as DARPA would be the first to admit, it "pursues research and technology where risk and payoff are both very high and where success may provide dramatic advances for traditional military roles and missions." In pursuing revolutionary advances, it specifically chooses to eschew an approach that will result merely in an "evolutionary improvement upon existing state-of-the-art." The topic of cyber insects was discussed in a report [5] by Princeton researchers in *Biology Letters*. Their team collected data on the migration patterns of dragonflies (*Anax junius*) by attaching (with superglue and eye-lash adhesive) battery-powered transmitters, weighing a mere 1/3 of a gram, to the insects and recording their flight data with a receiver located on a plane. Their hopes for the future of the project were *relatively* modest: "The dream scenario would be to get a satellite to pick up the signals from these transmitters." Compared with that, DARPA's hoped-for road seems to be (as they used to sing during World War I) "a long way to Tipperary."

REFERENCES

[1] The origin of the expression "*on a wing and a prayer*" is traced at: http://www. worldwidewords.org/qa/qa-ona2.htm (accessed October 22, 2015).

[2] The DARPA pre-solicitation notice (2006).

[3] R. Bansal, "Moths rush in where angels fear to tread," *IEEE Antennas and Propagation Magazine*, vol. 42, no. 2, p. 84, April 2000.

[4] G. Kitchener, "Pentagon Plans Cyber-Insect Army," *BBC* [Online]. Available: http://news.bbc.co.uk/2/hi/americas/4808342.stm (accessed October 22, 2015).

[5] "Tiny Tags Trace Dragonfly Paths," *BBC* [Online]. Available: http://news.bbc.co. uk/2/hi/science/nature/4759615.stm (accessed October 22, 2015).

(The original version of the column appeared in "Microwave surfing," *IEEE Microwave Magazine*, vol. 7, no. 5, pp. 28–30, October 2006.)

NOTE

1. To learn more about MEMS, see for example: https://www.mems-exchange.org/ MEMS/what-is.html (accessed May 5, 2016).

7.6 ELF COMMUNICATION: AN OBITUARY

The music in my heart I bore

Long after it was heard no more.

—W. Wordsworth (1770–1850)

These plaintive lines from Wordsworth's poem *The Solitary Reaper* came floating into my mind as I read the headline "Navy pulls plug on Project ELF" [1]. On September 30, 2004, at the end of the previous government fiscal year, the US Navy essentially terminated its $400 million Project ELF, when it silenced the twin ELF transmitters located in Michigan and Wisconsin.

The need for sending messages to deeply submerged ballistic-missile-carrying submarines was evident by the early 1960s, as the following quote from Captain Beach [2] makes clear:

"...If a Polaris submarine was to patrol on station with sixteen nuclear-tipped missiles on board, it was essential that positive, sure, national control be maintained over her operations. This was common sense, reinforced by a few other things such as Acts of Congress (including the original Atomic Energy Act), directives of the National Security Council, decisions of the Joint Chiefs of Staff, and flat-out orders from the Chief of Naval Operations..."

ELF (30–300 Hz) communication was found to be the solution to this problem since ELF waves can propagate around the world within the spherical waveguide formed by the earth and the ionosphere with minimal attenuation (around 1 dB per 1000 km [3]) and can penetrate seawater to useful depths with a relatively low attenuation (0.3 dB/m at 76 Hz [4]). It is worth recalling that the ELF communication channel is one-way only: from the shore to the submerged submarine. (As Collin [3] has explained, a signal coming from the submarine suffers attenuation on its way to the shore where it is easily overwhelmed by unattenuated atmospheric noise.) Even this low-bandwidth one-way ELF communication faced a tremendous technical hurdle in its implementation: when the wavelength is measured in thousands of kilometers, any practical transmitting antenna is electrically very small and, therefore, suffers from poor radiation efficiency.

To meet the challenge of transmitting a useful amount of ELF signal, the US Navy developed two huge transmitting stations in the Chequamegon National Forest near Clam Lake in northern Wisconsin and in Upper Michigan's Escanaba State Forest [5]. Each horizontal transmitting antenna employed miles of wires strung on hundreds of 40-feet poles. (This was a drastically scaled down version of the original 1960s plan, which would have involved a radiation-hardened grid of buried cable, thousands of miles long, and hundreds of transmitters.) The receiving antenna on the submarine took the form of a long insulated cable towed behind the submarine [4].

During their 15-year existence, the two transmitter locations were frequently targeted by peace activists and environmentalists. For the record, the Navy spent about $25 million on research and studies into public and environmental safety and found no problems [5]. Under political pressure to "re-evaluate its priorities," the Navy finally decided that it would no longer operate the two transmitters, which cost about $13 million a year. Even the associated infrastructure in Wisconsin and

Michigan was to be taken down over the next couple of years. The Navy now has to make do with its VLF transmitters located around the world for communicating with its submerged submarines. For those engaged in the development and implementation of the ELF communication system, an era suddenly came to an end.

REFERENCES

[1] "Navy Pulls Plug on Project ELF," *The Chief Engineer* [Online]. Available: http://chiefengineer.org/?p=1785 (accessed October 22, 2015).

[2] E. L. Beach, "ELF: to communicate with a submerged submarine," *Defense Electronics*, pp. 54–61, April 1980.

[3] R. E. Collin, *Antennas and Radio Wave Propagation*, McGraw Hill, 1985.

[4] D. F. Rivera and R. Bansal, "Towed antennas for US submarine communications: a historical perspective," *IEEE Antennas and Propagation Magazine*, vol. 46, no. 1, pp. 23–36, February 2004.

[5] R. Imrie, "Navy to Shut Down Sub Radio Transmitters," *USA Today* [Online]. Available: http://usatoday30.usatoday.com/tech/news/2004-09-26-sub-radio-offair_x.htm (accessed May 5, 2016).

(The original version of the column appeared in "AP-S turnstile," *IEEE Antennas and Propagation Magazine*, vol. 46, no. 6, p. 124, December 2004.)

NOTES

1. For a quick overview of submarine communication systems, see for example: http://www.globalsecurity.org/military/systems/ship/sub-comm.htm (accessed May 5, 2016).

2. Textbook resources:
   (i) W. H. Hayt and J. A. Buck, *Engineering Electromagnetics*, 8th ed., McGraw-Hill, New York, 2012. Electromagnetic wave propagation in seawater is discussed in Chapter 11.
   (ii) F. T. Ulaby and U. Ravaioli, *Fundamentals of Applied Electromagnetics*, 7th ed., Prentice Hall, Upper Saddle River, NJ, 2015. Electromagnetic wave propagation in seawater is discussed in Chapter 7.

7.7 CATCHING UP WITH PROFESSOR SCARRY

In Reference [1], I wrote about Elaine Scarry, a Harvard English (!) professor, who had put forward a theory of how electromagnetic interference (EMI) might have caused the explosion and crash of TWA Flight 800 on July 17, 1996 off Long Island, New York. After Professor Scarry published a long article "The Fall of TWA 800: The Possibility of Electromagnetic Interference" in the April 9, 1998 issue of *The New York Review of Books*, the National Transportation Safety Board (NTSB) "allocated several hundred thousand dollars for fresh research into EMI" and "Scarry's work was cited on the first page of a NASA study of EMI commissioned by TWA 800 investigators" [2].

Even though the NTSB ultimately ruled out EMI as a factor in the TWA disaster, Professor Scarry's fascination with EMI did not stop there. She subsequently published three more articles in *The New York Review of Books* [3], extending her EMI hypothesis to Swissair Flight 111, which went down near Nova Scotia after an electrical fire in September 1998, and EgyptAir Flight 990, whose October 1999 crash near Nantucket was widely attributed to the actions of a suicidal co-pilot. "As an explanation of any of these plane crashes—let alone all three—Scarry's theory is generally considered extremely unlikely" [2], but conspiracy theorists can delight is the "unnerving coincidences" [4] between SwissAir 111 and the TWA 800 flights. Both flights suffered unidentified electrical catastrophes, both took place during a week when extensive military exercises including Navy EP3 patrol aircrafts were under way in the area, both followed the same flight path prior to the reported problems, and (Lincoln/Kennedy assassination buffs would love this) both took off on a Wednesday at exactly 8:19 p.m. [4]. (Truth in advertising requires the recording of these other coincidences: Elaine Scarry received her Ph.D. from the University of Connecticut and now teaches at Harvard; the author of this book received his Ph.D. from Harvard and now teaches at the University of Connecticut!)

Perhaps more interesting than the plausibility of Scarry's theory is the question of why an English professor should be messing around with a specialized technical issue such as EMI in the first place? As I noted in Reference [1], the explanation lies in Scarry's interest in "cross-disciplinary" studies, which she described (to a CNN reporter) as "looking at certain questions to see how they occur across different fields or disciplines" such as law, medicine, and science. Scarry wants to use her skills in literary criticism "to solve social problems and save lives." She asserts, "There is nothing about being an English professor that exempts you from the normal obligations of citizenship. In fact, you have an increased obligation, because you know how to do research." Her Harvard colleague, Stephen Greenblatt, describes being in a car driven by her: "She took in each road sign, pondering all their possible meanings... She feels you need to read the whole world this way." The trouble is that, while this sort of all-encompassing approach to literary works readily wins acclaim, when applied to three separate aircraft accidents, it evokes skepticism (at least among the scientists). In the words of David Evans, the managing editor of *Air Safety Week*, "Coincidences, however compelling, do not add up to causality. The fact that the planes took off at the same minute the same night of the week and took the same flight path doesn't mean anything" [2].

Professor Scarry's speculations would no doubt have found a readier audience in the make-believe world of TV. According to an article in *Forbes* [5], *Dark Angel*, the sci-fi hit on Fox television, was set in 2019, 10 years after a nuclear electromagnetic pulse (NEMP) from a terrorist blast 50 miles off the East Coast had disabled all satellites, cable systems, computer databases, and (it appears) the entire financial infrastructure. I eagerly await Professor Scarry's deconstruction of the show.

REFERENCES

[1] R. Bansal, "AP-S turnstile: deconstructing the TWA crash," *IEEE Antennas and Propagation Magazine*, August 1998.

[2] E. Eakin, "Professor Scarry Has a Theory," *The New York Times Magazine*, pp. 78–81, November 19, 2000.

[3] *The New York Review of Books.* Full text of articles is available within the searchable archives of the journal at: http://www.nybooks.com/contributors/elaine-scarry/ (accessed May 5, 2016).

[4] "Could an EMC problem have brought down SwissAir 111?" *Conformity*, p. 6, December 2000.

[5] K. Blakeley, "Dark side of the script," *Forbes*, p. 63, January 22, 2001.

(The original version of the column appeared in "AP-S turnstile," *IEEE Antennas and Propagation Magazine*, vol. 43, no. 1, p. 122, February 2001.)

NOTE

1. A good starting point for finding technical material related to EMI and EMC is the website of the IEEE EMC Society: http://www.emcs.org/ (accessed May 5, 2016).

7.8 CRIMINAL INTERFERENCE

Reference [1] referred to *Dark Angel*, a sci-fi show on Fox television, set in 2019, 10 years after a NEMP from a terrorist blast 50 miles off the East Coast has disabled all satellites, cable systems, computer databases, and (it appears) the entire financial infrastructure. Similar scenes of chaos created by an electromagnetic bomb were conjured up by Ian Sample in a *New Scientist* article [2]. While (in the minds of most scientists) the threat of widespread havoc wreaked by NEMP has receded significantly since the end of the Cold War, the potential of employing EMI for criminal purposes is not as easily dismissed. Manuel Wink, at the time with the Defense Materiel Administration (Stockholm), was quoted by *Compliance Engineering* [3]:

> *"Our high-tech society depends heavily on systems that are vulnerable to electromagnetic high-power transient phenomena. Although the threat from criminal and terrorist activities is low to moderate today, it is thought that the threat will increase with time. Risk also increases with vulnerability. Therefore, we must closely follow developments in the field."*

Interestingly, at its General Assembly in Toronto in 1999, the URSI Council adopted the following Resolution on Criminal Activities using Electromagnetic Tools [4]:

"The URSI Council

Considering

(a) At the URSI General Assembly of 1984 a resolution was adopted on the adverse effects of a High Altitude Electromagnetic Pulse due to a Nuclear Explosion.

(b) The present resolution is intended to draw the attention of the scientific community to the effects of criminal activities using electromagnetic tools. This kind of action can be defined as an intentional malicious generation of electromagnetic energy introducing noise or signals into electric and electronic systems, thus disrupting, confusing or damaging these systems for terrorist or criminal purposes.

(c) Criminal activities using electromagnetic tools is an outgrowth of more familiar disciplines: Electromagnetic Compatibility (EMC) and EMI. In this case, however, the terrorist produces the offending currents or radiation intentionally. Accidental radiation can cause severe and inopportune damage to electronics, so those fields or more severe field levels can certainly also be intentionally impressed on vulnerable equipment. The electromagnetic compatibility community must be prepared to deal with new threats as they emerge.

This resolution is intended to make people aware of:

- *the existence of criminal activities using electromagnetic tools and associated phenomena.*
- *the fact that criminal activities using electromagnetic tools can be undertaken covertly and anonymously and that physical boundaries such as fences and walls can be penetrated by electromagnetic fields.*

- *the potential serious nature of the effects of criminal activities using electromagnetic tools on the infrastructure and important functions in society such as transportation, communication, security, and medicine.*
- *that in consequence, the possible disruption on the life, health and economic activities of nations could have a major consequence.*

It should be noted that the International Electrotechnical Commission (IEC) under Subcommittee 77C is developing a program to protect systems against these new EM threats.

Resolves

That URSI should recommend to the scientific community in general and the EMC community in particular to take into account this threat and to undertake the following actions:

1. Perform additional research pertaining to criminal activities using electromagnetic tools in order to establish appropriate levels of vulnerability.
2. Investigate techniques for appropriate protection against criminal activities using electromagnetic tools and to provide methods that can be used to protect the public from the damage that can be done to the infrastructure by terrorists.
3. Develop high-quality testing and assessment methods to evaluate system performance in these special electromagnetic environments.
4. Provide reasonable data regarding the formulation of standards of protection and support the standardization work which is in progress."

So far the response of the industry to the potential threat of criminal EMI activities has been slow in coming. William Radasky, at the time Chair of the International Electrochemical Commission (IEC) Committee 77, Subcommittee 77C, put it this way [3]:

"*We have in IEC a few people from industry—computer manufacturers and so forth—who are starting to become sensitive to this issue because they sell computers and they're worried about the security of those computers. There's a lot of technical interest from the research side, but in terms of organizations, IEC is really the only one that's working on this.*"

Since large clusters of servers have become a critical part of our economy, their reliability is clearly an important concern. A paper [5] presented at the 2001 EOS/ESD Symposium reported the results of an ongoing study of the EMI environment of server installations and potential threats to equipment. The authors noted that the threat posed by impulsive EMI to data stored in servers was as great or greater than the breach of physical security.

To provide some perspective on the levels of EMI that pose a threat, we can start with the European EMC Directive, which calls for immunity to field strengths of 3 V/m for residential equipment and 10 V/m for industrial equipment. According to Radasky, intentional EMI at a level of 100–200 V/m at a moderate distance can be produced with equipment purchased at a local Radio Shack or by using surplus military radar units. It should be noted that the threat of malicious EMI is not limited to radiated fields. As research at the Russian Academy of Sciences Institute for High Energy Densities has demonstrated, injection of disturbances into power lines outside a building can easily conduct damaging high voltages to computers inside the building [3].

Fortunately, protecting sensitive equipment against intentional EMI is not as hard for manufacturers as hardening it against NEMP. A little extra immunity, in the form of shielding or filtering, can be an effective countermeasure. Radasky noted [3]:

> *"It doesn't have to be 100 dB. Twenty dB may be enough. This factor of 10 makes it prohibitive for someone to cause a problem, because they have to be 10 times closer to cause the same field inside the shield."*

REFERENCES

[1] R. Bansal, "AP-S Turnstile: catching up with Professor Scarry," AP-S *Magazine*, vol. 43, no. 1, pp. 122–123, February 2001.

[2] I. Sample, "Wave of Destruction," *New Scientist* [Online]. Available: https://www.newscientist.com/article/dn698-wave-of-destruction/ (accessed October 22, 2015).

[3] "The New Cold War: Defending Against Criminal EMI," *Compliance Engineering*, pp. 12–18, May/June 2001.

[4] URSI Resolution on criminal EMI adopted at the 1999 General Assembly in Toronto.

[5] "The EMI/ESD Environment of Large Server Installations," *Conformity*, pp. 38ff, October 2001.

(The original version of the column appeared in "AP-S turnstile," *IEEE Antennas and Propagation Magazine*, vol. 43, no. 6, pp. 134–135, December 2001.)

NOTES

1. Reference [1] is included in this book as Section 7.7.
2. A good starting point for finding technical material related to EMI and EMC is the website of the IEEE EMC Society: http://www.emcs.org/ (accessed May 5, 2016).

7.9 WIRELESS NETWORKS: AN ELECTRONIC BATTLEFIELD?

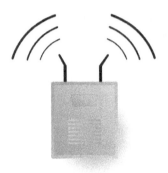

Dealing with covert surveillance and jamming may seem business as usual to microwave engineers working in the defense sector, but the same issues will appear rather remote to their colleagues in the wireless business. However, as the following two reports suggest, wireless network designers may need to use their own counter-measures to avoid being ambushed on the "electronic battlefield."

I Spy...

As the price of wireless technology has dropped, small wireless networks connecting computers within corporate buildings have proliferated. According to *The Wall Street Journal* [1], some 3.3 million wireless devices were shipped worldwide in 2000 and that number was expected to continue to grow. After all, a small network costs only a couple of hundred dollars, is relatively simple to set up, and does not require any messy cables snaking through the walls. The downside is that, unless the network administrators take steps to secure the network, the flow of information across the network is readily available to anyone snooping around. To make that point, Peter Shipley (a security consultant) and Matt Peterson (a wireless buff) drove around Silicon Valley, armed with a 2.45 GHz Yagi antenna hooked up to a laptop [1]. This modern equivalent of the Don Quixote and Sancho Panza team managed to tilt at windmills *successfully*: Shipley and Peterson routinely tracked emails and files flitting back and forth over unprotected networks at Silicon Valley companies. The software that makes a wireless network secure is readily available but, as Shipley and Petersen found, many wireless network users hadn't bothered to turn it on. As John Drewry of 3Com Corp. told the *Journal*, "Security is often an afterthought. A lot of education needs to happen." Sometimes a "rogue network" is set up by a few engineers at a company to share a printer without the knowledge of the corporate computer system administrators; such a network can render the whole system vulnerable to snooping. According to Shipley and Peterson, one doesn't have to drive right to a building to tap into the network. They had plans to demonstrate that, with the help of a special amplifier, they could carry out their surveillance of downtown networks from the hills above San Francisco. Caveat wireless network user!

Junk Bands

Who doesn't love a little multitasking? You don't want to stop using your spread-spectrum cordless telephone while keeping an eye on that cup of coffee in the microwave oven. But as a series of reports [2–4] warns, wireless communication (including the latest Bluetooth systems) may be jammed by perfectly legal RF emissions from a microwave oven operating nearby.

First a bit of background. Microwave ovens operate at 2450 MHz within one of the so-called ISM (industrial, scientific, and medical) bands. ISM equipment can operate at any power level in these bands and their RF emissions are restricted only by the international safety limits. Since some ISM equipment generates significant RF noise, it is not surprising that the three popular ISM bands (915, 2450, and 5800 MHz) are often dubbed "junk bands." The FCC rules allow non-ISM users to use these bands (e.g., for communication devices) provided they are willing to accept interference that may be caused by the operation of authorized ISM equipment [2]. Despite the potential of RF interference, these bands are attractive to the wireless community since systems built to operate in these bands can be used worldwide because of the universality of the ISM bands. Bluetooth systems are an example of low-power, short-distance, spread-spectrum communication devices that operate within the 2450 MHz band and which, because of their low power, come within the FCC's Part 15 rules and do not require a license to operate.

Experimental work [2] has shown that the operation of a nearby microwave oven can seriously degrade the performance of a spread-spectrum cordless telephone operating at 2450 MHz. Manufacturers of such telephones have been content with including a warning in the operating manual along the lines, "If you are near a microwave oven, noise may be heard at the receiver. Move away from it and closer to the base unit." However, if these ISM-band wireless communication devices deliver on the tremendous growth potential projected for them, there are likely to be serious compatibility issues in all kinds of locations using microwave ovens (and, by extension, others ISM devices such as RF lighting). Some solutions to mitigate this interference have been proposed [5] but not yet fully tested. Since both the sources of interference (microwave ovens, RF lighting) and the affected devices (wireless communication devices operating in ISM bands) meet the current FCC guidelines, it will be interesting to watch how this conflict plays out down the line. In the meantime, you may want to heat your coffee *before* powering up your Bluetooth gizmo.

REFERENCES

[1] L. Gomes, "Silicon Valley's Open Secrets," *The Wall Street Journal*, vol. 27, April 2001.
[2] C. Buffler and P. Risman, "Compatibility issues between bluetooth and high power systems in the ISM bands," *Microwave Journal*, pp. 126–134, July 2000.
[3] M. Lazarus, "ISM vs. spread spectrum—avoiding the FCC," *Microwave Journal*, pp. 116–122, October 2000.
[4] J. Osepchuk, private communication, April 12, 2001.

[5] P. Neelakanta and J. Sivaraks, "A novel method to mitigate microwave oven dictated EMI on bluetooth communications," *Microwave Journal*, pp. 70–88, July 2001.

(The original version of the column appeared in "Microwave surfing," *IEEE Microwave Magazine*, vol. 2, no. 4, pp. 32–34, December 2001.)

NOTES

1. A good starting point for finding technical material related to EMI and EMC is the website of the IEEE EMC Society: http://www.emcs.org/ (accessed May 5, 2016).
2. For more information about RF devices operating in the "junk bands," see for example: http://www.arrl.org/part-15-radio-frequency-devices (accessed May 5, 2016).

DID YOU KNOW?

A FUN QUIZ (VII)

1. For its 45th anniversary issue in 2006, *Microwaves & RF* created a list of 45 "Microwave Legends" (people, things, places) nominated and voted on by the magazine staff as well as the readers. Since then five new legends are inducted each year. An up-to-date list is available at the *Microwaves &RF* website. Who among the following has *not* yet made the list?

    (a) Maxwell

    (b) Faraday

    (c) Hertz

    (d) None of the above

2. Which of the following does *not* appear on The National Academy of Engineering (NAE) list of the 20 greatest engineering achievements of the twentieth century?

    (a) Radio and television

    (b) Radar

    (c) Laser and fiber optics

    (d) None of the above

3. Which of the following does *not* appear on The National Academy of Engineering (NAE) list of the 14 grand engineering challenges of the twenty-first century?

    (a) Develop magnetically levitated public transportation systems

    (b) Provide energy from fusion

    (c) Secure cyberspace

    (d) None of the above

4. LaserMotive, a start-up company from the Seattle area, won a $900,000 prize from NASA in 2009 for building a miniature prototype of a

    (a) Rocket using a fuel cell

    (b) Replacement for the space shuttle

    (c) Space elevator

    (d) None of the above

5. Researchers at Intellectual Ventures (Bellevue, WA) were working on _____ based systems to shoot down mosquitoes in the fight against malaria.

   (a) RF

   (b) Microwave

   (c) Laser

   (d) None of the above

6. The well-known Beloit College Mindset List is an annual effort (since 1998) to identify the worldview of 18-year olds. Which one of the following statements does *not* appear on the list compiled in the fall of 2009 (representing the frame of reference of the class of 2013)?

   (a) Wireless hot spots have always been around

   (b) Amateur radio operators have never needed to know Morse Code

   (c) Text has always been hyper

   (d) None of the above

7. An appendix in the 1928 third edition of *Practical Radio* by Moyer and Wostrel lists highlights of early radio history. Which of the following does *not* appear on the list?

   (a) Coherer (an appliance for detecting EM waves)

   (b) Self-heterodyne circuit

   (c) Coaxial cable

   (d) None of the above

8. The RFID transponders for "cyborgs" (insects carrying electronics) in a DARPA project used _____ power.

   (a) Nuclear

   (b) Fuel-cell

   (c) Photo-voltaic

   (d) None of the above

9. Who is considered the "father" of the cell phone?

   (a) Alexander Graham Bell

   (b) Marty Cooper

   (c) Al Gore

   (d) None of the above

10. Military satellite signals in the valleys of the mountainous regions of Afghanistan were often relayed by _____.

    (a) Blimps

    (b) Drones

    (c) Microwave towers

    (d) None of the above

ANSWERS

1. (b) Faraday

Source: "Microwave Legends," at http://mwrf.com/community/microwave-legends (accessed May 5, 2016).

2. (b) Radar

Source: R. Lucky, "Engineering achievements: the two lists," *IEEE Spectrum*, p. 23, November 2009.

3. (a) Develop magnetically levitated public transportation systems

Source: R. Lucky, "Engineering achievements: the two lists," *IEEE Spectrum*, p. 23, November 2009.

4. (c) Space elevator

Source: K. Chang, "Winner in Contest Involving Space Elevator," *The New York Times*, November 8, 2009.

5. (c) Laser

Source: "Zap!" *The Economist Technology Quarterly*, p. 10, June 6, 2009.

6. (a) Wireless hot spots have always been around

Source: "Beloit College Mindset List" at https://www.beloit.edu/mindset/previouslists/2013/ (accessed May 5, 2016).

7. (c) Coaxial cable

Source: "Design notes: a bit of radio history," *High Frequency Electronics*, p. 64, April 2009.

8. (a) Nuclear

Source: The report on DARPA's nuclear-powered (radio isotope) cyborgs is in the December 2009 issue of the *IEEE Spectrum* available at http://spectrum.ieee.org/semiconductors/devices/nuclearpowered-transponder-for-cyborg-insect (accessed May 5, 2016).

9. (b) Marty Cooper

Source: "Father of the Cell Phone," *The Economist Technology Quarterly*, pp. 30–31, June 6, 2009.

10. (a) Blimps

Source: "Spies in the Sky," *The Economist Technology Quarterly*, p. 6, June 6, 2009.

(The original version of the quiz appeared in "AP-S turnstile," *IEEE Antennas and Propagation Magazine*, vol. 52, no. 1, pp. 197–212, February 2010.)

CHAPTER 8

DOMESTIC AND INDUSTRIAL APPLICATIONS

"Oh that a man's reach should exceed his grasp, or what's a Heaven for"
—Robert Browning (1812–1889)

8.1 BLOWIN' IN THE WIND

Q. What mode of transport did Marty McFly, the teenager who traveled from 1985 to October 21, 2015 in the 1989 Hollywood film "Back to the Future II," use to get away in a hurry from a gang of bullies?

A. Of course, a hoverboard!

From ER to E.T.: How Electromagnetic Technologies Are Changing Our Lives, First Edition. Rajeev Bansal.
© 2017 by The Institute of Electrical and Electronic Engineers, Inc. Published 2017 by John Wiley & Sons, Inc.

Well, October 21, 2015 has come and gone, so what about that hoverboard? I am glad you asked.

Bob Gale, the writer behind the "Back to the Future" trilogy, imagined [1] that the "hoverboard floats on a magnetic field similar to magnetic levitation trains." As far as magnetic levitation trains are concerned, as I wrote in a column [2] a few years ago, any traveler to China can experience the exhilarating magnetic levitation technology aboard the Shanghai Maglev Train to the Pudong International airport (top speed in excess of 250 mph). However, for most people, magnetic levitation has remained a scientific curiosity, good enough for wowing students in a freshman physics laboratory with an inexpensive lab set up, but with few realizable practical applications. But, enough about trains. You were asking about the practicality of Maglev hoverboards.

Well, as widely reported in the media [3, 4], a recent youtube video featuring the "Lexus hoverboard" has been viewed nearly 10 million times. "Footage released by the firm [Lexus] showed skateboarders testing the hoverboard with varying degrees of success. The film was recorded on a *specially constructed* skate park near Barcelona in Spain [3]." I italicized the words *specially constructed* to point out that the concrete surface of the skate park in the video is a tease and there are presumably magnets located underneath the surface. The skateboard itself contains a superconductor cooled to $-321°F$ by liquid nitrogen (hence the wisps of "smoke" coming out of the sides). Mike Norman, Director of the Materials Science Division at Argonne National Lab, explains [4]: "There's interaction [Meissner effect] between the superconductor and the magnet that repels the force of gravity and allows the thing to levitate. That's why it can't be pure concrete in the video; there has to be something magnetic there as well."

At this point, you may feel discouraged that such superconducting hoverboards will not be at your local sporting goods stores until our streets have suitable magnetic tracks built underneath. But, if you permit me another flight of fancy, I would like you to ponder over the implications of the following development. A group of physicists in Spain has reportedly [5] created a magnetic "wormhole" in the laboratory "that is capable of transporting a magnetic field from one point in space to another." Imagine the thrilling future where our hero could be riding a superconducting hoverboard through a magnetic wormhole. Do I hear a casting call for Back to the Future IV?

REFERENCES

[1] C. Dougherty, "Hoverboard? Still in the Future," *The New York Times*, October 21, 2014 [Online]. Available: http://www.nytimes.com/2014/10/21/technology/hoverboard-still-in-the-future.html?_r=0 (accessed August 21, 2015).

[2] R. Bansal, "AP-S turnstile: a eureka moment," *IEEE Antennas and Propagation Magazine*, vol. 53, no. 3, pp 174–175, June 2011. DOI: 10.1109/MAP.2011.6028445

[3] "Levitating Magnetic Hoverboard Unveiled," *BBC News* [Online]. Available: http://www.bbc.com/news/technology-33785285 (accessed August 21, 2015).

[4] B. Barrett, "How That Lexus Hoverboard Actually Works," *Wired* [Online]. Available: http://www.wired.com/2015/06/lexus-hoverboard-slide/ (accessed August 21, 2015).

[5] E. Cartlidge, Physicists create a magnetic wormhole, *PhysicsWorld* [Online]. Available: http://physicsworld.com/cws/article/news/2015/aug/20/physicists-create-a-magnetic-wormhole-in-the-lab (accessed August 21, 2015).

(The original version of the column appeared in "AP-S turnstile," *IEEE Antennas and Propagation Magazine*, vol. 57, no. 6, p. 150, December 2015.)

NOTES

1. Reference [2] is included in this book as Section 1.7.
2. For an explanation of the Meissner effect, see for example: http://lrrpublic.cli.det.nsw.edu.au/lrrSecure/Sites/Web/physics_explorer/physics/lo/superc_12/superc_12_02.htm (accessed May 6, 2016).
3. To learn more about wormholes, see for example: http://www.space.com/20881-wormholes.html (accessed May 6, 2016).

8.2 SHARING THE ROAD

Storrs, where the main campus of the University of Connecticut is located, is a small rural community. Yet, as I walk to work from my home, I rarely see bicyclists on the road. I recall once asking a northern European graduate student why he did not ride his bike to work and he told me that the bicyclists around here did not feel safe sharing the road with cars (except where there are physically separate bike lanes). Unlike Europe, drivers here often do not "look out" for or slow down for bicyclists. At least, one car company with roots in Northern Europe, has been doing something to address this problem [1].

At the 83rd Geneva International Motor Show in Switzerland [2], Volvo unveiled a bicyclist detection system, which should soon be available on many of its models. An enhanced version of the pedestrian detection system, which Volvo launched in 2010, it uses [2] a radar in the car's grille as well as a camera located behind the windshield. The technology is supposed to be able to detect multiple targets simultaneously, including the possibility of a bicyclist suddenly swerving in front of the car. According to Volvo, if a collision appears to be imminent, not only will an alarm sound to alert the driver but the car's brakes will also be automatically engaged. While Volvo owners with the previous pedestrian detection systems installed in their cars will be able to upgrade their system software to take advantage of the new bicyclist detection feature, anyone interested in the system in a new car will have to order it *before* the car is manufactured in the factory. The system was supposed to add nearly $3000 to the cost of the car as part of a package of optional features.

According to data from the Statistical Abstract of the United States: 2012 [3], some 700 bicyclists and nearly 5000 pedestrians are killed annually in motor vehicle accidents. Therefore, any efforts on the part of the auto manufacturers to increase the safety of people on the road are welcome. However, British Cycling, the United Kingdom's governing body for bicyclists, told the BBC [2]:

> "*While we obviously welcome any safety measures that can be built into vehicles, people shouldn't be relying on technology to keep them and other road users safe. What would make much more of a difference is making cyclist awareness a mandatory part of the driving test. British Cycling will continue to campaign for this as well as the establishment of a prominent, national cyclist awareness campaign similar to that we've seen for motorcyclists.*"

For its part, Volvo has been considering other safety enhancements as well. In 2013, it released its first car equipped with an airbag under the hood, which would inflate and cover a good portion of the windshield if the sensors in the front bumper detect that the car has come in contact with a person. Longer term, company engineers have been conducting experiments in safari parks on animal behavior to devise technologies that would be useful for detecting horses and deer in the path of a car. Now, that is something that would get serious attention from a lot of Connecticut car drivers, where road accidents involving deer are increasingly a major problem.

REFERENCES

[1] "Volvo Unveils Cyclist Alert-and-Brake Car System," *BBC News* [Online]. Available: http://www.bbc.co.uk/news/technology-21688765 (accessed October 29, 2015).

[2] R. Hamilton "Hot Cars at the 83rd Geneva International Motor Show," *The Baltimore Sun* [Online]. Available: http://darkroom.baltimoresun.com/2013/03/hot-cars-at-the-83rd-geneva-international/}1 (accessed October 29, 2015).

[3] "Transportation: Motor Vehicle Accidents and Fatalities," *Statistical Abstract of the United States: 2012* [Online]. Available: http://www.census.gov/library/publications/2011/compendia/statab/131ed/transportation.html (accessed October 29, 2015).

(The original version of the column appeared in "AP-S turnstile," *IEEE Antennas and Propagation Magazine*, vol. 55, no. 2, p. 190, April 2013.)

NOTES

1. To learn more about automotive radar systems, see for example: http://spectrum.ieee.org/transportation/advanced-cars/longdistance-car-radar (accessed May 6, 2016).

2. Textbook resources:

F. T. Ulaby and U. Ravaioli, *Fundamentals of Applied Electromagnetics*, 7th ed., Prentice Hall, Upper Saddle River, NJ, 2015. The basic operation of a radar is presented in Chapter 10.

8.3 RARE NO MORE?

Here is a little quiz:

Dysprosium and neodymium

(a) *Are rare-earth metals*
(b) *Are strongly magnetic*
(c) *Are mined mostly in China*
(d) *All of the above*

Yes, indeed, dysprosium and neodymium are all of the above but one may well wonder why anyone would give the matter a second thought. Of course, if one were in the business of manufacturing electric and hybrid cars or offshore wind turbines, one would readily appreciate their importance because permanent magnets based on these rare earths have been a key component of the electric motors used in such applications [1–5]. According to a report [2] in *MIT Technology Review*, "the motor of Toyota's Prius, for example, uses about a kilogram of rare earths while offshore wind turbines can require hundreds of kilograms each" because alloys of rare-earth metals, for example, neodymium with iron and boron, are "four to five times as strong by weight as permanent magnets made from any other material." It also cited [2] in an estimate from the US Department of Energy (DOE) "that widespread use of electric-drive vehicles and offshore wind farms could cause shortages of these metals by 2015."

Ironically, despite their name, rare-earth metals are relatively abundant in Earth's crust. However, they [2] "are usually found mixed together in deposits that often contain radioactive elements as well—and separating the metals requires costly processes that produce a stew of toxic pollutants.... The [extraction process also] generates a lot of salt: when the Mountain Pass mine [in California] was running at full capacity in the 1990s, it produced as much as 850 gallons of salty wastewater every minute, every day of the year." The mine ran into financial and environmental problems and was shut down in 2002 but, with the recent surge in the demand for (and the price of) rare earths, it has come back to life as Molycorp Minerals and now claims to have a more sustainable extraction process [2].

The US DOE, through its high risk/high reward arm ARPA-E [3], has supported 14 projects that comprised ARPA-E's REACT program, short for "Rare Earth Alternatives in Critical Technologies." The goal was to develop "cost-effective alternatives to rare earths... The REACT projects will identify low-cost and abundant replacement materials for rare earths while encouraging existing technologies to use them more efficiently."

Moving beyond material science, some European and Japanese researchers are exploring electric motor designs that do away with permanent magnets altogether. One such design is a switched reluctance motor. "The idea behind it is over 100 years old, but making a practical high-performance version suitable for vehicles has not been possible until recently. A combination of new motor designs and the advent of powerful, fast-switching semiconductor chips, which can be used to build more sophisticated versions of the electronic control systems required to operate a reluctance motor, is giving those motors a new spin" [1].

While it may be too early to predict which of these emerging approaches (permanent magnets based on new materials or new motor designs that dispense with permanent magnets) will eventually see commercial success in the growing fields of electric vehicles and wind turbines, it is hearting to see that sustainable alternatives are receiving so much government and private investment worldwide.

REFERENCES

[1] "Reluctant Heroes," *The Economist* [Online]. Available: http://www.economist.com/news/science-and-technology/21566613-electric-motor-does-not-need-expensive-rare-earth-magnets-reluctant-heroes (accessed October 29, 2015).

[2] K. Bourzac, "The Rare-Earth Crisis," *MIT Technology Review* [Online]. Available: http://www.technologyreview.com/featuredstory/423730/the-rare-earth-crisis/ (accessed October 29, 2015).

[3] "Rare Earth Alternatives in Critical Technologies (REACT)," *ARPA-E* website [Online]. Available: http://www.arpa-e.energy.gov/?q=arpa-e-programs/react (accessed October 29, 2015).

[4] "An Impossible Dream?" *The Economist* [Online]. Available: http://www.economist.com/blogs/babbage/2012/03/rare-earths-and-high-performance-magnets (accessed October 29, 2015).

[5] B. Reddall and J. Gordon, "Analysis: Search for Rare Earth Substitutes Gathers Pace," *Reuters* [Online]. Available: http://www.reuters.com/article/2012/06/22/us-rareearths-alternatives-idUSBRE85L0YB20120622 (accessed October 29, 2015).

(The original version of the column appeared in "AP-S turnstile," *IEEE Antennas and Propagation Magazine*, vol. 54, no. 6, p. 179, December 2012.)

NOTES

1. To learn more about the basics of electric motors, see for example: http://electronics. howstuffworks.com/motor.htm (accessed May 6, 2016).
2. Textbook resources:
   (i) W. H. Hayt and J. A. Buck, *Engineering Electromagnetics*, 8th ed., McGraw-Hill, New York, 2012. Magnetic materials are discussed in Chapter 8.
   (ii) F. T. Ulaby and U. Ravaioli, *Fundamentals of Applied Electromagnetics*, 7th ed., Prentice Hall, Upper Saddle River, NJ, 2015. Magnetic materials are discussed in Chapter 5.

8.4 LOCAL HEATING?

After the energy crisis of the 1970s, Robert V. Pound (1919–2010), then professor of physics at Harvard University and later a recipient of the National Medal of Science [1], put forward a modest proposal [2] in 1980. He felt that conventional means of heating houses and buildings were very inefficient since they involved not only keeping the inhabitants warm but also, in the process, heating all the air inside the room (imagine a multi-story atrium). Furthermore, given the time required to bring a room to a comfortable temperature, spaces were kept warm whether or not they were occupied. Professor Pound's idea was to think of each room in a building as a microwave cavity, where a magnetron discreetly hidden behind an opening in one of the walls, could be turned on to warm the inhabitants instantly without wasting energy on heating the air. *Voilà*!

Dr. Charles R. Buffler (1934–2001) and a colleague decided to put the idea to test [2] with themselves as the willing subjects. They used a 10-foot cubical metallic chamber connected to a microwave generator that could supply up to 500 W, put a couple of chairs inside, and turned the "microwave oven" on. According to Dr. Buffler [2], the success of the experiment led him to conjecture that "in large quantities, microwave home room heating systems could be sold for under one hundred dollars, thereby dramatically reducing heating energy costs." But he conceded that "the utilization of microwaves for whole body heating of humans probably awaits decades of discussion and research as well as the advent of one or more prolonged energy crises."

While Professor Pound's proposal to use microwave heating of people never really got off the ground, his insight that heating the inhabitants in a room is much more energy efficient than heating the whole space, has found a new life recently in the "local warming" project [3] conceived by Dr. Carlo Ratti, Director of the SENSEable City Laboratory at MIT. Dr. Ratti, an architect and engineer by training, apparently got his inspiration [3] "while sitting outside a restaurant being warmed by one of those tall, mushroom-shaped infra-red heaters." He figured that, by using optics controlled by servo motors, heat from infrared lamps (or arrays of infrared LEDs) could be directed at people as they went about their business in a building foyer. The system also uses Wi-Fi-based sensors to track people. The concept was demonstrated both under the porch near MIT's main entrance and at the 14th International Architecture Exhibition held in Venice in fall 2014. Dr. Ratti estimates that, in large spaces like building atriums, "local heating" may save up to 90% of the traditional heating costs.

The "local warming" project can be coupled with a smart-phone-based app so that each person can get the LEDs targeted at him to deliver the desired level of heat. Dr. Ratti says [3] that "it's almost like having your personal sun following you around." This observation brought to my mind the "kangri" [4], a traditional personal heating device used in Kashmir, which consists of an earthen pot for burning charcoal, carried around one's neck in a wicker basket. And, unlike the MIT project, this one works even where there is no Wi-Fi.

REFERENCES

[1] Robert Vivian Pound [Online]. Available: http://news.harvard.edu/gazette/story/2012/10/robert-vivian-pound/ (accessed October 29, 2015).

[2] C. R. Buffler, "Whole body microwave heating of humans and livestock," *Harvard Graduate Society Newsletter*, pp. 11–13, Summer 1988.

[3] "In the Moment of the Heat," *The Economist* [Online]. Available: http://www.economist.com/news/technology-quarterly/21615065-one-way-keep-warm-heat-people-rather-expending-energy-heating (accessed October 29, 2015).

[4] B. Kinu, "Kashmiri Kangri - An Age-Old Device for Keeping Warm" [Online]. Available: http://www.demotix.com/news/920229/kashmiri-kangri-age-old-device-keeping-warm}media-919991 (accessed October 29, 2015).

(The original version of the column appeared in "AP-S turnstile," *IEEE Antennas and Propagation Magazine*, vol. 56, no. 5, pp. 200–201, October 2014.)

NOTE

1. 🔲 Which of the following can produce significant heating?
   (a) Microwaves
   (b) Ultrasound waves

(c) Infrared radiation

(d) All of the above

 (d) All of the above

The main point to remember here is that in terms of thermal effects, high-intensity microwaves are not fundamentally different from other more conventional sources of heat. For example, therapeutic hyperthermia (controlled heating of tissues in a hospital setting) can be produced by a hot water bath or ultrasound or RF/micro-wave energy.

Source: R. Bansal, "Pop quiz: EMF and your health," *IEEE Potentials*, pp. 3–4, August/September 1997.

8.5 COMING SOON TO A WAL-MART NEAR YOU

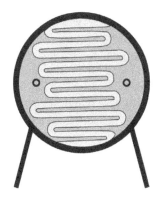

About 10 years ago, at a Retail Systems conference in Chicago, a top Wal-Mart executive gave a presentation that attracted a standing room only crowd. No, the executive did not promise $99.99 high-definition TVs for the holiday shopping season. Nor did she talk about Wal-Mart's position on immigration and labor issues. Instead, Wal-Mart CIO Linda Dillman was there to announce [1] that, within a few years, the company would require pallet-level and carton-level radio tagging from its top 100 suppliers. Welcome to the exciting world of RFID!

RFID Technology

According to the website [2] of AIM (Automatic Identification and Mobility), radio-frequency identification (RFID) technology is an automatic way to collect product, place, time, or transaction data quickly and easily without human intervention or error. An RFID system consists of a reader, which uses an antenna to transmit radio energy to interrogate a transponder (radio tag, RFID card). The transponder does not have a battery but rather receives its energy from the incoming RF signal. The energy is used to extract data stored in an integrated circuit (chip) and send it to the reader, from where it can be fed to a computer for processing.

A Brief History

J. Landt's paper "Shrouds of Time: the History of RFID" (available online at [3]), includes the following milestones in the history of RFID technology:

- The 1940s: H. Stockman's paper, "Communication by means of reflected power," *Proceedings of the IRE*, pp. 1196–1204, October 1948. (Stockman concluded: "Evidently, considerable research and development work has to be done before the remaining basic problems in reflected-power communication are solved, and before the field of useful applications is explored.")
- The 1950s: D. B. Harris's "Radio Transmission Systems with Modulatable Passive Responder," US Patent #2,927,321.
- The 1960s: J. P. Vinding's invention, "Interrogator-Responder Identification System," 1967. Early commercial products used 1-bit tags, whereby the presence or absence of the tagged item could be detected as part of an anti-theft electronic article surveillance (EAS) technology.
- The 1970s: A. Koelle, S. Depp, and R. Freyman, "Short-Range Radio-Telemetry for Electronic Identification Using Modulated Backscatter," 1975. Large companies such as Raytheon ("Raytag" in 1973), RCA (F. Sterzer's electronic license plate in 1977), and GE were engaged in development work.
- The 1980s: The first RFID-based operational toll collection system was used in Norway in 1987.
- The 1990s: Widespread use of electronic toll collection systems, for example, "E-Z Pass." The development of single-chip microwave RFID tags.

Promise of RFID Technology

As market analyst Christopher Boone said [4], "we are at an incredibly early stage of this technology and what it is actually capable of doing." With the ability to track everything from crates of disposable razors to individual peanut butter jars on the store shelves, RFID technology offers the potential of "real-time supply chain visibility." Promoters of RFID technology feel [4] that "RF tags are to this decade what the Internet was to the 1990s – a promise of radical change in the way business is done."

Hurdles

However, before the full potential of RFID technology can be realized, several hurdles need to be overcome [4]:

Reliability: RF tags have yet to reach the 99% reliability rate enjoyed by UPC tags because of RF noise in operational conditions (e.g., from nylon conveyor belts).

Cost: Many analysts believe that before the widespread adoption of RFID technology becomes a reality, the price of tags (in large volumes) will have to drop below 5 cents a tag from its current 20–30 cents apiece. There is also the upfront cost of converting the manufacturing and retail systems from UPC to RFID.

Lack of standards: RFID systems of today employ many competing mutually incompatible protocols.

Security: Early efforts to use RFID technology for inventory management have raised concerns that tags will be used to develop consumer profiles. Civil liberties advocates similarly fear that RFID technology may be used by the government to monitor people. In this context, it should be noted that RFID technology is short range (a few centimeters to a few meters); it cannot be used for satellite-based surveillance [1].

Conclusion

As with any emerging technology, the hurdles faced by RFID technology will gradually diminish (prices will drop with increasing volumes, uniform standards will emerge with market consolidation, etc.). Of course, many of the small companies marketing RFID technology today may not have the resources to survive long enough to benefit from the larger RFID market down the line, but as large clients like Wal-Mart publicly embrace the technology, the future of RFID looks increasingly brighter.

REFERENCES

[1] "The commercial market," *Microwave Journal*, p. 65, September 2003.

[2] The website for AIM [Online]. Available: http://www.aimglobal.org/?page=About_AIM (accessed October 29, 2015).

[3] The website for Transcore [Online]. Available: https://www.transcore.com/literature (accessed October 29, 2015).

[4] "Radio Tags Face Technical Hurdles, Deadlines," *The Age* [Online]. Available: http://www.theage.com.au/articles/2003/11/12/1068329599083.html?from=storyrhs (accessed October 29, 2015).

(The original version of the column appeared in "AP-S turnstile," *IEEE Antennas and Propagation Magazine*, vol. 45, no. 6, pp. 105–106, December 2003.)

NOTES

1. Sections 8.6 also discusses RFIDs.
2. For the "RF" aspects of an RFID, see for example:
    D. Dobkin, *The RF in RFID,* 2nd ed., Newnes, 2012.

8.6 HAS YOUR CAT GONE PHISHING?

Burt Kaliski, the then Director of the EMC Innovation Network, once observed: *"Depending on your perspective, RFID technology is either a dream come true or your worst, paranoid nightmare."* There has been a well-known concern [1, 2] about the threat to privacy in a future "Big Brother" society saturated with RFID tags. An award-winning paper "Is Your Cat infected with a Computer virus?" [3], presented at PERCOM'06, added yet another dimension to the fears stirred by RFID technology. The paper illustrates the general feasibility of RFID malware (worms, viruses, phishing, etc.) by presenting the design of the first ever RFID virus.

An RFID system has two parts: a (usually portable) reader and a tiny transponder (radio tag, RFID chip), which is embedded in or attached to the tracked object (such as a piece of baggage on an airport conveyor belt or a pet). Researchers from Vrije Universiteit in Amsterdam demonstrated in Reference [3] how these tiny RFID tags could be used to spread malicious computer code. Since the tags have a very limited memory (typically less than 1024 bits), it had been generally assumed that they were unsuitable vectors for introducing viruses into computers connected to RFID readers [4].

To skeptics who wondered how a resource-limited tag can launch an attack, the research team from Vrije Universiteit responded that an RFID attack "requires more ingenuity than resources." In particular, "the manipulation of less than 1 Kbits of on-tag RFID data can exploit security holes in RFID middleware, subverting its security, and perhaps even compromising the entire computer, or the entire network!" The researchers mentioned buffer overflows, code insertion, and SQL injection as examples of exploits that RFID tags could perform. Their paper included a detailed description of an RFID virus implementation that was small enough to fit on a low-cost tag with 127 characters [3].

To dramatize their point, Rieback et al. [3] evoke a scene at a veterinarian's clinic. At first the vet's RFID-based pet identification system starts behaving oddly by displaying, for example, erroneous address information for the pets. Sometime later, the system starts erasing data from the pets' (rewritable) RFID tags. In the climactic scene, the veterinarian's computer screen freezes and displays the chilling message "All your pet are belong to us" [*sic*]. In another example, the authors describe how a tag infected with a worm and attached to a piece of baggage can not only infect other baggage [tags] at the airport but also, "on arrival at other airports, these cases will be scanned again and within 24 hours, hundreds of airports throughout the world could be infected" [4].

The RFID industry's reaction to the Dutch researchers' claims was generally one of skepticism. An article in *RFIDUpdate* [5] noted: "A key premise of the researchers' assertions is that the bits and bytes stored on RFID tags would be interpreted by [RFID] readers as executable instructions. The reality is that tag contents are never interpreted as executable code; they are interpreted only as simple raw data, like numbers. For an RFID system to interpret tag data otherwise would require a poor, insecure design that breaks the most basic and obvious rules of system engineering." In fairness to the Dutch research team, it should be noted that, after demonstrating how to exploit RFID middleware systems to launch an attack, the Dutch paper lists various steps that concerned parties can take to protect their computer systems against RFID-induced malware. The March 2007 cover of the *IEEE Spectrum* featured an RFID enthusiast who had had RFID tags implanted in his palms for opening doors, unlocking his car, and logging into his computer. The Rieback et al. paper will probably give other early adopters of RFID technology pause.

REFERENCES

[1] R. Bansal, "AP-S turnstile: coming soon to a Wal-Mart near you," *IEEE Antennas and Propagation Magazine*, vol. 45, no. 6, pp. 105–106, December 2003.

[2] R. Bansal, "Microwave surfing: now you see it and now you don't," *IEEE Microwave Magazine*, vol. 5, no. 4, pp. 32–34, December 2004.

[3] M. Rieback, B. Crispo, and A. Tanenbaum, "Is Your Cat infected with a Computer virus?" *Proceedings of the Fourth Annual IEEE International Conference on Pervasive Computing and Communication* (PERCOM'06), March 13–17, 2006, Pisa, Italy. Available online at http://www.rfidvirus.org/ (accessed October 30, 2015)

[4] W. Knight, "RFID Worm Created in the Lab," *New Scientist*, March 15, 2006. Available online at http://www.rfidvirus.org/ (accessed October 30, 2015)

[5] "The Industry Reacts to RFID Virus Research," *RFIDUpdate*, March 20th 2006.

(The original version of the column appeared in "AP-S turnstile," *IEEE Antennas and Propagation Magazine*, vol. 50, no. 3, p. 142, June 2008.)

NOTES

1. Reference [1] has been included in this book as Section 8.5.
2. For the "RF" aspects of an RFID, see for example:
   D. Dobkin, *The RF in RFID*, 2nd ed., Newnes, 2012.

8.7 THE FUTURE OF WIRELESS CHARGING

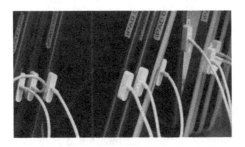

Ah, the freedom of walking around with a sleek ultra-portable computer, surfing the web, while also chatting with a friend on a mobile phone, all through invisible wireless networks! Unfortunately, sooner or later, the computer and phone batteries run out of juice and one is forced to search for their specific power adapters and tether them to a power strip. If only we could charge these devices wirelessly!

Some years ago, a British company Splashpower [1, 2] took a step in the right direction when it designed a system that enables one to recharge portable devices simply by laying them on a "SplashPad." While the pad a device rests on has to be plugged into the wall, there is no need to hunt for any special adapters and the portable device is, indeed, receiving its power "wirelessly." Michael Faraday, the British father of electromagnetic induction, would have been impressed, although he once said [3]: "Nothing is too wonderful to be true if it be consistent with the laws of nature." The trouble is Splashpower was declared bankrupt in 2008 without releasing any commercial products.

Splashpower may have been ahead of its time, but the technology of electromagnetic inductive coupling remains attractive to other players in the wireless-charging arena. The trick is to persuade manufacturers to abandon their proprietary adapters and incorporate inductive charging modules into their devices. This is where the Wireless Power Consortium [4] comes into play. Formed in late 2008, the international Consortium is a "group of leading manufacturers in a wide range of industries that understand the untapped potential of wireless charging." The group focuses on a wireless power technology that transmits power only to a product which is in close proximity of the charging station. In this way, the transmitter can focus the energy on the receiver safely and with high efficiency. The consortium promotes an open standard for all. With a universal wireless power charging standard, electronic products and charging stations using the same standard will recognize each other and be mutually compatible. The 200+ members of the Consortium include Dell, Samsung, Texas Instruments, and Fulton Innovation (the company that acquired the assets of Splashpower).

The dream of long-range wireless power transmission (WPT), pursued by Nikola Tesla in the nineteenth century, and more recently by the US DOE and the National Aeronautics and Space Administration (NASA) (remember solar power satellite (SPS) systems [5] proposed in the 1970s?), is also alive and well. Powercast, a Pittsburgh-based company, has developed a range of wireless-charging products, based on RF energy harvesting, that promise to deliver [3, 6] "milliwatts over meters" and "watts over meters" for charging low-power lighting and wireless sensors among potential applications. Unfortunately, the RF power needed to charge cell phones and laptops over moderate distances may not pass muster with the regulators because of potential human health hazards [3]. PowerBeam [7], a silicon valley startup, which replaces radio waves with low-power laser beams for wireless charging may also encounter regulatory hurdles [3].

So is wireless charging ever going to become a practical reality? The answer is a cautious yes, but in the meantime do not forget to pack all your adapters.

REFERENCES

[1] R. Bansal, "AP-S turnstile: cutting the cord," *IEEE Antennas and Propagation Magazine*, vol. 49, no. 1, p. 150, February 2007.

[2] "One Charging Pad Could Power Up All Gadgets," *New Scientist* [Online]. Available: https://www.newscientist.com/article/dn6891-one-charging-pad-could-power-up-all-gadgets/ (accessed October 30, 2015).

[3] "Adaptor Die," *The Economist* [Online]. Available: http://www.economist.com/node/13174387 (accessed October 30, 2015).

[4] Wireless Power Consortium website [Online]. Available: http://www.wirelesspower consortium.com/ (accessed May 6, 2016).

[5] J. McSpadden and J. Mankins, "Space solar power programs and microwave wireless power transmission technology," *IEEE Microwave Magazine*, pp. 46–57, December 2002.

[6] Powercast company website [Online]. Available: http://powercastco.com/ (accessed October 30, 2015).

[7] PowerBeam company website [Online]. Available: http://www.powerbeaminc.com/ (accessed October 30, 2015).

(The original version of the column appeared in "AP-S turnstile," *IEEE Antennas and Propagation Magazine*, vol. 51, no. 2, p. 153, April 2009.)

8.8 ELECTROPOLLUTION OR SUSTAINABLE ENERGY?

electropollution *n.* Nonionizing electromagnetic radiation propagated through the atmosphere by broadcast towers, radar installations, and microwave appliances, and the magnetic fields surrounding electrical appliances and power lines, which is believed to have polluting effects on people and the environment; also called electromagnetic smog [1].

As the above dictionary definition makes it clear, manmade electromagnetic radiation often gets a bad rap as a form of pollution or smog. Since engineers have been turning all kinds of garbage into useful energy (e.g., biodiesel from used cooking oil), it was only a matter of time before someone tapped into the energy latent in the radio waves all around us. For example, a 2010 story in *The New York Times* [2] described innovative work in this area by Matt Reynolds and Jochen Teizer, who had been interested in developing a hard hat ("SmartHat") that would alert the worker through an audio signal when dangerous construction equipment happened to be nearby. Their SmartHat prototype included a microprocessor and a beeper mounted under the visor of a hard hat. The electronic circuit worked without any batteries. Rather, the microprocessor harvested the needed electrical energy from the ambient radio-frequency (RF) fields. In this case, the RF fields were a byproduct of the wireless network transmitters that were mounted on backhoes and bulldozers to keep track of their locations. So, on the one hand, the SmartHat microprocessor monitored the strengths of the RF signals from the construction equipment to alert the hat's wearer when the equipment gets too close. On the other, microprocessor also derived the needed dc power for its operation from the surrounding fields.

In another application reported in the same *The New York Times* story, a team from Intel and the University of Washington created a temperature and humidity sensor, which drew its power form the fields created by a TV broadcast antenna located a few miles away. These are all, of course, low-power applications with the needed power in microwatts. However, that is not as limiting as it may appear at first glance; a typical solar-powered calculator needs only about 5 microwatts [2].

What distinguishes the applications discussed above from the RF energy harvesters used by companies such as Powercast [3] is that the current generation of Powercast devices makes use of a *dedicated* RF transmitting unit rather than relying on the *ambient* RF fields. However, even Powercast foresees that the technology will evolve from a paired system with the need for a dedicated transmitter to a single-sided system with the ability to fully capture radio waves emitted from existing and commonly used ambient RF energy sources, such as mobile base stations, TV and radio transmitters, microwave radios, and mobile phones.

All this would not have come as a surprise to Marconi, an early pioneer in wireless transmission. In a 1912 interview published in the *Technical World Magazine* [4], he prophesied: "Within the next two generations we shall have not only wireless telegraphy and telephony, but also wireless transmission of all power for individual and corporate use, wireless heating and light, and wireless fertilizing of fields."

REFERENCES

[1] "Electropollution," *Dictionary.com* [Online]. Available: http://dictionary.reference.com/browse/electropollution?&o=100074&s=t (accessed October 29, 2015).

[2] A. Eisenberg, "Bye-Bye Batteries: Radio Waves as a Low-Power Source," *The New York Times*, July 18, 2010.

[3] Powercast company website [Online]. Available: http://powercastco.com/ (accessed October 29, 2015).

[4] I. Narodny, "Marconi's Plans for the World," *Technical World Magazine*, October 1912, pp. 145–150 [Online]. Available: http://earlyradiohistory.us/1912mar.htm (accessed October 29, 2015).

(The original version of the column appeared in "AP-S turnstile," *IEEE Antennas and Propagation Magazine*, vol. 52, no. 4, p. 134, August 2010.)

NOTE

1. Section 8.7 also discusses techniques for the wireless transmission of power.

DID YOU KNOW?

A FUN QUIZ (VIII)

For this Quiz, I have chosen all the questions from the various feature articles that appeared in the *IEEE Microwave Magazine* during 2008.

1. *"It is widely accepted that _____ are essential to the success of an R&D department working predominantly on passive components."*
   (a) Empirical models
   (b) 3D numerical simulations of EM fields
   (c) Building hardware and measuring s-parameters
   (d) None of the above

2. *"_____ has today become a critical part of the microwave design cycle."*
   (a) Numerical simulation
   (b) Electromagnetics (EM)
   (c) Nonlinear circuit theory
   (d) None of the above

3. *"The most common filter technology for both the RF BAW and SAW filters is the same: the _____ type."*
   (a) Ladder
   (b) π-network
   (c) T-network
   (d) None of the above

4. *"The second type of chipless RFID transponders that are known to exist is based on _____ materials, which are tiny particles of chemicals."*
   (a) Isometric
   (b) Paramagnetic
   (c) Nanometric
   (d) None of the above

5. *"For digital communications, the sustainable _____ is a commonly used benchmark for measuring system performance."*
   (a) Data rate
   (b) BER

(**c**) Bandwidth

(**d**) None of the above

6. *"Wireless propagation cannot be improved in the same manner as propagation in _____ was improved."*

(**a**) Coaxial cables

(**b**) Microwave waveguides

(**c**) Glass optical fiber

(**d**) None of the above

7. *"An estimate that _____ of these pins [in a chip] are dedicated to signal lines is not unreasonable."*

(**a**) 20–30%

(**b**) 50–60%

(**c**) 80–90%

(**d**) None of the above

8. *"Single-mode optical fiber dominates terrestrial long-haul communications because it has losses as low as _____."*

(**a**) 0.02 dB/km

(**b**) 0.2 dB/km

(**c**) 2 dB/km

(**d**) None of the above

9. *"The _____ configuration is quickly becoming the device pair of choice for the RF [power amplifier] industry."*

(**a**) Doherty

(**b**) Darlington

(**c**) CMOS

(**d**) None of the above

10. *"Radio communications in the past century have relied primarily on _____ to modulate and demodulate signals for wireless transmissions."*

(**a**) Vacuum tubes

(**b**) Semiconductor devices

(**c**) Nonlinear devices

(**d**) None of the above

ANSWERS

1. (**b**) 3D numerical simulations of EM fields
   Source: T. Weiland, M. Timm, and I. Munteanu, "A practical guide to 3-D simulation", *IEEE Microwave Magazine*, vol. 9, pp. 62–75, December 2008.

2. (b) Electromagnetics (EM)
Source: J. Rautio, "Shortening the design cycle," *IEEE Microwave Magazine*, vol. 9, pp. 86–96, December 2008.

3. (a) Ladder
Source: F. Bi and B. Barber, "Bulk acoustic wave RF technology," *IEEE Microwave Magazine*, vol. 9, pp. 65–80, October 2008.

4. (c) Nanometric
Source: S. Preradovic, N. Karmakar, and I. Balbin, "RFID transponders," *IEEE Microwave Magazine*, vol. 9, pp. 90–103, October 2008.

5. (a) Data rate
Source: S. Chia, T. Gill, L. Ibbetson, D. Lister, A. Pollard, R. Irmer, et al., "3 G evolution," *IEEE Microwave Magazine*, vol. 9, pp. 52–63, August 2008.

6. (c) Glass optical fiber
Source: D. Cox and H. Lee, "Physical relationships," *IEEE Microwave Magazine*, vol. 9, pp. 89–94, August 2008.

7. (a) 20–30%
Source: T. G. Ruttan, B. Grossman, A. Ferrero, V. Teppati, J. Martens, "Multiport VNA measurement," *IEEE Microwave Magazine*, vol. 9, pp. 56–69, June 2008.

8. (b) 0.2 dB/km
Source: S. Iezekiel, "Measurement of microwave behavior in optical links," *IEEE Microwave Magazine*, vol. 9, pp. 100–120, June 2008.

9. (a) Doherty
Source: R. Sweeney, "Practical magic," *IEEE Microwave Magazine*, vol. 9, pp. 73–82, April 2008.

10. (c) Nonlinear devices
Source: R. G. Bosisio, Y. Y. Zhao, X. Y. Xu, S. Abielmona, E. Moldovan, Y. S. Xu, et al., "New-wave radio," *IEEE Microwave Magazine*, vol. 9, pp. 89–100, February 2008.

(The original version of the quiz appeared in "Microwave surfing," *IEEE Microwave Magazine*, vol. 10, no. 4, pp. 30–62, June 2009.)

CHAPTER 9

COMMUNICATION SYSTEMS

Small is Beautiful

—E. F. Schumacher (1911–1977)

9.1 SMALL IS BEAUTIFUL

In times of congressional discord over the *right* size of the government, it is natural to think that the title of this section refers to British Economist E. F. Schumacher classic collection of essays entitled *Small Is Beautiful: A Study of Economics As If People Mattered* [1]. However, my concern here is not with the size of our budget but with the size of antennas. During my long professional career as a professor and sometime consultant in applied electromagnetics, I have been asked numerous times why the

From ER to E.T.: How Electromagnetic Technologies Are Changing Our Lives, First Edition. Rajeev Bansal.
© 2017 by The Institute of Electrical and Electronic Engineers, Inc. Published 2017 by John Wiley & Sons, Inc.

antenna community does not avail itself of the wonders of the micro/nano-electronic revolution. After all, if transistor designers can constantly miniaturize their devices in response to the relentless demands of Moore's law, surely we antenna designers can also do something to drag ourselves out of the era of ridiculous rabbit ears. Well, I answer, we have. Mobile phones work better than ever and you can't even see the antennas. But I know that the truth will out. From rooftop antenna dishes to space probes bristling with ungainly communication hardware, we have not really succeeded in transcending the wavelength-based limits imposed on antenna designers by Maxwell's equations. Yet, within those immutable constraints, antenna engineers continue to "squeeze" more antenna performance into a given volume. Consider the case of antennas [2, 3] proposed for miniature satellites.

CubeSats are spacecraft the size of a shoebox and weighing a few pounds that can "often piggyback on the launches of larger space missions" [2] and can be used to put scientific equipment into Earth orbit relatively inexpensively. Of course, their small size has so far limited CubeSats to small antennas (see, e.g., [4]) suitable for low data rates and limited communication range (low orbits: 300–1000 km from the Earth). There is a great deal of interest among scientists to use CubeSats for interplanetary missions. This would, of course, require larger antennas. A Jet Propulsion Laboratory (JPL) group led by Alessandra Babuscia [3] has been investigating an antenna formed from a mylar (film thickness around 50 micrometers) balloon that can be fitted into a 10 cm^3 volume for the launch and later expanded to a width of 1 m [2].

While inflatable antennas have been used in space missions before, they previously required cumbersome systems using pressure valves. The new antenna design relies upon a small amount of benzoic-acid powder stored within the balloon, which sublimates in the reduced-pressure environment of outer space and results in the inflation of the balloon. (The sublimation concept was used by NASA in the 1960s for some balloon satellites; in the current version it is being used for the communication antenna [2].)

While the antenna inflation has been tested in a vacuum chamber and radiation properties characterized (presented at the 2014 IEEE Aerospace Conference), research continues on evaluating the effect of penetration of the inflated antenna by micrometeorites in the space environment. Simulations suggest that, even with small leaks, antenna can remain inflated for several years and can transmit data 10 times faster and 7 times further than current CubeSat antennas [2]. The reaction of the system to collisions with bigger particles still needs to be investigated.

REFERENCES

[1] E. F. Schumacher, *Small Is Beautiful: A Study of Economics As If People Mattered*, Vintage, New Ed edition (1993).

[2] J. Dacey, "Inflatable Antenna Could Send Tiny Satellites Beyond Earth Orbit," *Physicsworld* [Online]. Available: http://physicsworld.com/cws/article/news/2013/sep/17/inflatable-antenna-could-send-tiny-satellites-beyond-earth-orbit (accessed December 17, 2015).

[3] A. Babuscia, B. Corbin, R. Jensen-Clem, M. Knapp, I. Sergeev, M. Van de Loo, et al., "CommCube 1 and 2: A CubeSat series of missions to enhance communication capabilities for CubeSat," *2013 IEEE Aerospace Conference Proceedings*, pp. 1–19. DOI: 10.1109/AERO.2013.6497128.

[4] R. M. Rodriguez-Osorio and E. F. Ramirez, "A hands-on education project: antenna design for inter-CubeSat communications," (Education Column), *IEEE Antennas and Propagation Magazine*, vol. 54, no. 5, pp. 211–224, October 2012.

(The original version of the column appeared in "AP-S turnstile," *IEEE Antennas and Propagation Magazine*, vol. 55, no. 5, pp. 168–169, October 2013.)

NOTES

1. To learn more about cubesats, see for example: http://www.cubesat.org/ (accessed May 9, 2016).
2. Textbook resources:
   (i) W. H. Hayt and J. A. Buck, *Engineering Electromagnetics*, 8th ed., McGraw-Hill, New York, 2012. Antennas are discussed in Chapter 14.
   (ii) F. T. Ulaby and U. Ravaioli, *Fundamentals of Applied Electromagnetics*, 7th ed., Prentice Hall, Upper Saddle River, NJ, 2015. Antennas are discussed in Chapter 9. Satellite communication systems are discussed in Chapter 10.

9.2 GIGABIT WI-FI

The year 2012 marked the 100th volume for the *Proceedings of the IEEE,* testifying to the relentless march of electrical and electronic technologies through the twentieth century. It also marked the second year for a new kid on the block: the *IEEE Transactions on Terahertz Science and Technology*. The terahertz (THz) band, spanning the electromagnetic spectrum from 300 GHz to 30 THz and mostly unregulated [2] around the world, used to represent a vast wilderness in the field of communications. There were good reasons for that: THz signals required cumbersome, expensive, and power-hungry devices for generation and detection [3], they suffered significant attenuation as they traveled through air, and there was no compelling need for the huge bandwidth offered by the band. While the physics underlying the absorption of THz waves by air molecules remains just as unforgiving as ever, making long-distance (several kilometers) wireless THz transmission impractical even with very high-gain antennas [1], the overall landscape has changed recently in favor of an expanding interest in the THz band for communication. (It may be noted in passing that the THz band is also being exploited for imaging in research, security, and biomedical applications.)

Let us start with the need for the wide communication bandwidths available in the THz band. It has been predicted that the traffic from wireless devices may soon exceed that from wired devices [1]. And, a lot of that traffic may well be over short distances, for example, for high-speed wireless links between servers in a data center, for transmitting media from an audio-visual (AV) rack to a high-definition TV, for super-fast wireless transfers between handheld devices [2]. In order for THz-based devices to capture that market, one would need transmitter and receiver components that are small (to fit into a smart phone), power-efficient (to run off rechargeable batteries), and, last but not least, relatively inexpensive. Researchers from the Tokyo Institute of Technology have brought these goals a step closer to reality.

The Japanese group demonstrated a 3 Gbps data transfer rate over a 542 GHz wireless connection, thereby doubling the previous record of 1.5 Gbps reported by chipmaker Rohm in November 2011 [2, 3]. These blistering speeds are limited to distances of the order of 10 m but that is a still a good-sized room. At the heart of the Japanese THz system lies a specially developed resonant tunneling diode (RTD) that acts as an oscillator and replaces traditional complex THz generators such as quantum cascade lasers. In retrospect, it is easy to see why a December 2011 conference, co-sponsored by IEEE-USA in cooperation with the Federal Communications Commission (FCC) Office of Engineering and Technology, was dubbed [2] "Terahertz Technology: The Next Frontier for Radio." A video presentation from the conference on the potential of THz communication technology by David Britz of AT&T is available at Reference [4]. Stay tuned.

REFERENCES

[1] T. Schneider, A. Wiatrek, M. Grigat, and R.-P. Braun, "Link budget analysis for terahertz communication," *IEEE Transactions on Terahertz Science and Technology*, vol. 2, no. 2, pp. 250–256, March 2012.

[2] J. Keene, "Japanese Researchers Break the Terahertz Wireless Transmission Speed Record," *The Verge* [Online]. Available: http://www.theverge.com/2012/5/16/3023676/terahertz-wireless-record-3gbps-tokyo-university (accessed December 17, 2015).

[3] "Milestone for Wi-Fi With 'T-Rays'," *BBC News* [Online]. Available: http://www.bbc.co.uk/news/science-environment-18072618 (accessed December 17, 2015).

[4] "Terahertz Technology: Terahertz Wireless Communication," *AT&T Tech Channel* [Online]. Available: http://techchannel.att.com/play-video.cfm/2012/1/19/Conference-TV-Terahertz-Technology:-Terahertz-Wireless-Communication1 (accessed December 17, 2015).

(The original version of the column appeared in "AP-S turnstile," *IEEE Antennas and Propagation Magazine*, vol. 54, no. 4, p. 226, August 2012.)

NOTE

1. The full range of the electromagnetic spectrum is displayed at, for example: http://earthsky.org/space/what-is-the-electromagnetic-spectrum (accessed May 9, 2016).

9.3 OPEN SPECTRUM: A TRAGEDY OF THE COMMONS?

"Picture a pasture open to all. It is to be expected that each herdsman will try to keep as many cattle as possible on the commons…The effects of overgrazing are shared by all herdsmen…Therein lies the tragedy" [1].

—G. Hardin in "The Tragedy of the Commons"

In the wake of the Titanic disaster, when nearby ships had not responded to the Titanic's distress signal, there were calls for government regulation of the electromagnetic spectrum ("the commons"). The Radio Act of 1912 required all broadcasters to register with the government. In response to lobbying by the broadcasters, the Federal Radio Commission was created in 1927, which led to the Federal Communications Commission (FCC) 7 years later. The goal was the exclusive licensing of well-separated bands ("individual fenced-in pastures") to broadcasters with a view to avoiding a cacophony of interfering signals ("the tragedy"). Since the primitive radio receivers of the period couldn't distinguish among different transmitters very well, this exclusive licensing of the EM spectrum seemed the right approach [2, 3].

For most of its existence, the FCC has been playing Santa Claus, giving away wide swaths of the electromagnetic spectrum *free* to users ranging from city governments (e.g., police, fire departments) to TV broadcasters. In the 1990s, the "prime" spectrum (30–3000 MHz) having been mostly assigned, Congress suddenly realized the market value of the available bandwidth and decided to auction a portion of it to the emerging cell phone industry for billions of dollars. It is, therefore, natural to believe that, at this point, the prime EM spectrum must be heavily congested. Interestingly, experiments conducted in Santa Rosa, CA [2] and Pittsburgh [4] revealed that except for some heavily congested slivers, for example, the "PCS" and the Wi-Fi bands, most of the prime spectrum was being used only sporadically. Also, many chunks of the spectrum have served very limited markets; for example, Pax TV (a UHF broadcast group rebranded as Ion Television in 2007), used to air infomercials for household gadgets for much of the day [2].

Has the time come to rethink the allocation of the EM spectrum as a "commons" available to all users? Wouldn't that lead to a "tragedy of the commons" (a chaotic situation preventing everyone from communicating reliably)? Radical as it may sound, at least two different approaches are already being tried to have a more open spectrum. The first is to allocate *specific* bands for unlicensed devices with only general rules to promote peaceful coexistence between different users. The 2.4 GHz and 5 GHz Wi-Fi bands are examples of the successful implementation of this strategy. The second approach is to "underlay" unlicensed technologies in *existing* bands, without disturbing licensed users. Underlay may be achieved by using a signal with a very low spectral power density. For example, the FCC has authorized the use of ultra-wideband (UWB) technology devices in this fashion [3].

Another potential enabling technology for the implementation of open spectrum is software/cognitive radio [5]. Joseph Mitola, who coined the terms software radio and cognitive radio discussed their development in this way [6]:

> "*Software radios are emerging as platforms for multiband multimode personal communications systems. Radio etiquette is the set of RF bands, air interfaces, protocols, and spatial and temporal patterns that moderate the use of the radio spectrum. Cognitive radio extends the software radio with radio-domain model-based reasoning about such etiquettes. Cognitive radio enhances the flexibility of personal services through a radio knowledge representation language.*"

While Edmond Thomas, the then chief of the FCC's Office of Engineering and Technology, conceded the potential benefits of cognitive radio in opening up the EM spectrum, he did not expect the FCC to open up the entire spectrum anytime soon [5]. For the time being, when it comes to spectrum management, a fear of the "tragedy of the commons" remains the prevailing ethos.

REFERENCES

[1] G. Hardin, "The tragedy of the commons," *Science*, vol. 162, no. 3859, pp. 1243–1248, 1968.

[2] S. Woolley, "Dead Air," *Forbes*, pp. 138–150, November 25, 2002.

[3] K. Werbach, "Open Spectrum: The New Wireless Paradigm," New America Foundation, Spectrum Series Working Paper #6, October 2002.

[4] J. Jackson, "Breakthrough technologies: using the airwaves more efficiently," *Washington Technology*, July 15, 2002.

[5] P. Rojas, "Thinking of Radio as Smart Enough to Live Without Rules," *The New York Times*, October 24, 2002.

[6] J. Mitola and G. Maguire, Jr., "Making software radios more personal," *IEEE Personal Communications*, vol. 6, no. 4, pp. 13–18, August 1999.

(The original version of the column appeared in "AP-S turnstile," *IEEE Antennas and Propagation Magazine*, vol. 44, no. 6, pp. 123–124, December 2002.)

NOTES

1. The full range of the electromagnetic spectrum is displayed at (for example): http://earthsky.org/space/what-is-the-electromagnetic-spectrum (accessed May 9, 2016).
2. To learn more about radio spectrum allocation in the United States, consult the FCC website at: https://www.fcc.gov/engineering-technology/policy-and-rules-division/general/radio-spectrum-allocation (accessed May 9, 2016).

9.4 NEAR-FIELD COMMUNICATION

In the early years of the twenty-first century, even as cell phones shrank in size while boasting an ever-increasing array of features, two things about them did not change much: they often sprouted a stubby antenna and, if one wanted to use a headset, one had to put up with an unwieldy wire connecting the headset and the phone. The antenna was eventually made to vanish inside the cell phone case by using an internal microstrip-style printed version. Thanks to a technology called near-field communication (NFC), one can also cut the cord between the phone and the headset *without* sacrificing audio quality or data security [1–4].

NFC Technology

According to the website for Aura Communications (now Freelinc) [3], while the concepts behind magnetic induction communication had been around for decades, Aura's engineers were the first to develop and implement practical solutions capturing the benefits of this technology. NFC communicates wirelessly by coupling a very low power quasi-static magnetic field at 13.56 MHz (one of the frequencies available worldwide for unlicensed industrial, scientific, and medical (ISM) applications). Such a field may be produced, for example, with an electrically small loop antenna ("magnetic dipole"). Polarization diversity is employed to provide nearly omnidirectional reception [3]. In analogy with the electric field of an electric dipole (think *duality*), the magnetic field of a magnetic dipole exhibits a $1/r^3$ dependence on distance in the near zone. Normally, this range-limiting rapid rate of decay would be a serious handicap, compared with the $1/r$ drop off of the far field of more familiar RF wireless technologies, for example, Bluetooth operating at 2.45 GHz. But in really short-range (1–2 m) applications such as the link between a cell phone or an MP3 player and a headset, this rapid fall off is exploited to provide each user with his own private "bubble" without having to worry about mutual interference among multiple users and permitting bandwidth reuse. (Theoretically, a wireless link based on quasi-electric field should work just as well, but the quality of such a link suffers greatly in the presence of commonly encountered conducting objects. Magnetic fields, on the other hand, are not affected by human bodies and non-magnetic objects in the vicinity [3].)

Advantages

The physics of quasi-static magnetic fields leads to a number of desirable features in devices equipped with the NFMI technology:

- **Lower Power Consumption:** Since signals are limited to a very short range (typically centimeters), NFC devices require very little power and may have up to a sixfold advantage in terms of battery power over Bluetooth-enabled devices [3]. One of the early commercial NFC products, the LibertyLink Docker, was claimed to be good for several hours of talk-time on a single AA battery [2].
- **Available Bandwidth:** Since NFC devices do not operate in the crowded 2.45 GHz band and since each user is "sealed" within his own bubble, frequency reuse is greatly facilitated for NFC devices. For streaming music applications such as MP3 players, a bandwidth of 384 kbits/s with a bit error rate (BER) of 10^{-5} may be needed to provide the equivalent of hard-wired service quality [4]. With NFC technology, that is easily achievable for multiple users in the same area.
- **Increased Reliability:** Because the magnetic near field falls off rapidly with distance, NFC devices do not have to contend with fading due to multipath. As a result, NFC devices offer a much more robust service quality compared with Bluetooth-type devices [3].

Conclusion

Many companies are offering modestly priced (less than $50) NFC products to consumers. In 2004, the then Aura CEO Kokinakis told *The New York Times* [1] in an interview, "I want to become the de facto standard for personal communications devices delivering voice and audio." That, of course, is a tall order, but the concept of using near-field magnetic fields for short-range communication certainly continues to show a lot of promise.

REFERENCES

[1] A. Krauss, "For Audio Players, A Chance to Cut the Cord," *The New York Times*, March 4, 2004.

[2] D. Wolfson, "The LibertyLink Docker Wireless Headset," a product review in *Computing Unplugged*, January 1, 2004.

[3] Aura Communications (now part of Freelinc) website [Online]. Available: http://www.freelinc.com/technology/ (accessed December 17, 2015).

[4] V. Palermo, "Near-Field Magnetic Comms Emerges," *Electronic Engineering Times*, November 3, 2003.

(The original version of the column appeared in "AP-S turnstile," *IEEE Antennas and Propagation Magazine*, vol. 46, no. 2, pp. 114–115, April 2004.)

NOTES

1. To learn more about near-field communication (NFC), see for example: http://www.nearfieldcommunication.org/

2. Textbook resources:
   (i) W. H. Hayt and J. A. Buck, *Engineering Electromagnetics*, 8th ed., McGraw-Hill, New York2012. Magnetic dipole antennas are discussed in Chapter 14.
   (ii) F. T. Ulaby and U. Ravaioli, *Fundamentals of Applied Electromagnetics*, 7th ed., Prentice Hall, Upper Saddle River, NJ, 2015. The field produced by a magnetic dipole is discussed in Chapter 5.

9.5 A NEW DIGITAL PHONE?

How excited would you be if someone offered you a new digital phone receiver that weighed 170 tons and had a data rate of 0.1 Hz? My guess is, probably, not too much. How about if the receiver had a *truly* universal reach, staying in touch under deep sea on planet earth or even somewhere out there in the galaxy in E.T. country? Well, now you may be giving it a second thought.

A paper published in *Modern Physics Letters A* [1] and discussed in several magazines [2–4] describes the first successful demonstration of just such a "phone." The phone is the brainchild of an international group of particle physicists working at Fermi National Accelerator Laboratory in Illinois and it uses neutrinos (yes, those chargeless, nearly massless elusive particles) for digital messaging. In the experiment reported in Reference [1], a series of accelerators at the lab produces an energetic beam of protons, which strikes a carbon target giving rise to pions, kaons, and many other exotic particles. These particles pass through a 675 m helium-filled "decay pipe," where most of the particles decay into neutrinos. The beam then passes through 240 m of rock (mostly shale) where all particles except neutrinos are absorbed. Since neutrinos interact extremely weakly with matter, they can travel immense distances and pass through all kinds of materials (including the earth and seawater) with little attenuation or defocusing. Of course, that also makes it a challenge to detect the neutrinos and, hence, the need for a massive "receiver."

The MINERvA detector used in the experiment [1] is located in an underground cavern about a kilometer from the source. The basic unit of the detector is [1] "a hexagonal plane assembled from parallel triangular scintillating strips. The full detector has 200 such planes, and a total weight of 170 tons." The neutrino beam has a transverse size of a few meters at the face of the detector. In the demonstration, most of the detected signal is from neutrino interactions (resulting in the production of muons) in the rock lying between the source and the detector. A smaller component of the signal comes from the neutrino interactions within the active region of the detector.

The neutrino beam is encoded by controlling the proton beam pulses at the source, using simple on–off keying scheme (with a "1" representing a beam pulse and a "0" the absence thereof). An intensity of 2.25×10^{13} protons per pulse results, on average, in a mere 0.81 event registered at the detector. In the study reported in Reference [1], the pulses were 8.1 μs wide and spaced 2.2 seconds apart. The first message transmitted repeatedly was the binary-coded word "NEUTRINO." The data received consisted of 3454 records spanning an interval of 142 minutes. "An overall data rate of about 0.1 Hz was realized, with an error rate of less than 1% for transmission of neutrinos through a few hundred meters of rock" [1].

Such neutrino-based communication systems have been proposed (but not demonstrated) for electromagnetically challenging environments such as a scenario involving a submerged submarine [5]. Of course, a neutrino detector will have to shrink a great deal in weight before that enters the realm of possibility. But, before we dismiss the concept of neutrino-based communication itself out-of-hand, we should think about what happened to Hedy Lamarr and her invention in 1942 of a frequency-hopping radio-controlled system for guiding torpedos. As I described in Reference [6], Hollywood actress Lamarr and her co-inventor George Antheil (a composer) had proposed a player-piano mechanism as *one* possible implementation of the frequency-hopping system. Antheil later wrote [6], "In our patent Hedy and I attempted to better elucidate our mechanism by explaining that certain parts of it worked like the fundamental mechanism of a player piano. Here, undoubtedly, we made our mistake. The reverend and brass-headed gentlemen in Washington who examined our invention read no further than the words 'player piano.' 'My God,' I can see them saying, 'we shall put a player piano in a torpedo.'… In 1962, three years after the Lamarr-Antheil patent had expired, ships equipped with secure military-communication systems, based on the frequency-hopping technique, were deployed during the Cuban missile crisis." So, even neutrino phones may someday see the light of day.

REFERENCES

[1] D. Stancil, P. Adamson, M. Alania, L. Aliaga, M. Andrews, C. Araujo Del Castillo, et al., "Demonstration of communication using neutrinos," *Modern Physics letters A*, vol. 27, no. 4, 2012. DOI: 10.1142/S0217732312500770

[2] "ET, Phone Home," *The Economist* [Online]. Available: http://www.economist.com/node/21550242 (accessed December 17, 2015).

[3] J. Aron, "Neutrinos Send Wireless Message Through the Earth," *New Scientist* [Online]. Available: http://www.newscientist.com/blogs/shortsharpscience/2012/03/neutrinos-send-wireless-messag.html (accessed December 17, 2015).

[4] R. Boyle, "For the First Time, a Message Sent With Neutrinos," *Popular Science* [Online]. Available: http://www.popsci.com/science/article/2012-03/first-time-neutrinos-send-message-through-bedrock (accessed December 17, 2015).

[5] J. Hsu, "Neutrinos May Someday Provide High-Speed Submarine Communication," *Popular Science* [Online]. Available: http://www.popsci.com/military-aviation-amp-space/article/2009-10/neutrinos-may-someday-provide-high-speed-submarine-communication (accessed December 17, 2015).

[6] R. Bansal, "He(a)dy stuff," *IEEE Antennas and Propagation Magazine*, vol. 39, no. 3, p. 100, June 1997.

(The original version of the column appeared in "AP-S turnstile," *IEEE Antennas and Propagation Magazine*, vol. 54, no. 2, pp. 182–183, April 2012.)

NOTE

1. Reference [6] is included in this book as Section 1.1.

9.6 ELECTRONIC COUNTERMEASURES

US Presidential candidates have to put up with a lot during the course of their campaigns but there is a limit to their tolerance. The Republican candidate (2008) Rudy Giuliani apparently reached his limit during a campaign speech before a crowd of 18,000 in Iowa when he heard a cell phone ring just beyond the stage. He stopped in the middle of his story, turned to the offending cell phone's owner, and said sarcastically, "It's okay; you can answer your cell phone. You won't interrupt me" [1].

Of course, you don't have to be a stressed out contender in a presidential primary to be annoyed by someone else's cell phone in a public space. It happened to a California architect taking his morning train to work when a young woman sitting next to him started "blabbing away using the word 'like' all the time." Instead of confronting his fellow passenger a la Giuliani, Andrew engaged in a bit of covert electronic countermeasure. He put a hand in his pocket and activated a device not much larger than a cigarette case. What happened next would have pleased even the techno-wizards who used to supply James Bond with all the cool gadgets his movies are famous for. The device in Andrew's shirt emitted a radio signal that disrupted cell phone communication in a 30-feet radius [2].

As many of the readers might have guessed, the technology behind cell phone jamming is pretty simple. The device broadcasts a radio signal in the frequency range reserved for cell phones, which seriously interferes with the cell phone signal, resulting in a "no network available" display on the cell phone screen [3]. The jammer is not target specific; all phones within the effective radius are silenced indiscriminately. This countermeasure does not require the brute power of a microwave oven to be effective; international websites selling such devices tout radiated power of 20 milliwatts.

As I noted in Reference [4], cell phone jamming devices were (and remain) illegal in the United States. The FCC website [5] even spells out the penalties for jamming:

> "The operation of transmitters designed to jam or block wireless communications is a violation of the Communications Act of 1934, as amended ('Act'). See 47 U.S.C. Sections 301, 302a, 333. The Act prohibits any person from willfully or maliciously interfering with the radio communications of any station licensed or authorized under the Act or operated by the U.S. government. 47 U.S.C. Section 333. The manufacture, importation, sale or offer for sale, including advertising, of devices designed to block or jam wireless transmissions is prohibited. 47 U.S.C. Section 302a(b). Parties in violation of these provisions may be subject to the penalties set out in 47 U.S.C. Sections 501-510. Fines for a first offense can range as high as $11,000 for each violation or imprisonment for up to one year, and the device used may also be seized and forfeited to the U.S. government."

Nonetheless overseas exporters have reported [2] an increasing demand for portable jammers with hundreds of units shipped to the United States every month. Cell phone service providers who not only paid billions of dollars to lease the cell phone frequencies from the government but continue to spend vast sums to maintain and expand their networks are not pleased. A Verizon spokesman told *The New York Times* [2]: "It's counterintuitive that when the demand is clear and strong from wireless consumers for improved cell coverage, that these kinds of devices are finding a market." Perhaps, what we need, is what Dan Briody, in an article [6] in *InfoWorld* called the "ten commandments of cell phone etiquette." For example, his commandment #2 reads: "Thou shalt not set thy ringer to play La Cucaracha every time thy phone rings. Or Beethoven's Fifth, or the Bee Gees, or any other annoying melody. Is it not enough that phones go off every other second? Now we have to listen to synthesized nonsense?" Unfortunately, it is hard to be optimistic that many people would heed Briody's sensible commandments. I recall the headline from a newspaper some years ago [7]: "Spanish King telling Chavez to 'shut up' becomes ringtone hit in Spain." Time to google "cell phone jammers" yet?

REFERENCES

[1] "Rudy and Other People's Cellphones," *The Atlantic: Daily Dish* [Online]. Available: http://andrewsullivan.theatlantic.com/the_daily_dish/2007/09/rudy-and-other-.html (accessed October 29, 2015).

[2] M. Richtel, "Devices Enforce Silence of Cellphones, Illegally," *The New York Times*, November 4, 2007.

[3] D. Bennahum, "Hope You Like Jamming, Too," *Slate* [Online]. Available: http://www.slate.com/id/2092059/ (accessed October 29, 2015).

[4] R. Bansal, "Microwave surfing: knock on wood," *IEEE Microwaves Magazine*, vol. 5, no. 1, pp. 38–40, March 2004.

[5] Current version of the applicable FCC regulations [Online]. Available: https://www.fcc.gov/encyclopedia/jammer-enforcement (accessed October 29, 2015).

[6] D. Briody, "The Ten Commandments of Cell Phone Etiquette" [Online]. Available: http://www.appleseeds.org/10-Commands_Cell-Phone.htm (accessed October 29, 2015).

[7] "Spanish King Telling Chavez to 'Shut Up' Becomes Ringtone Hit in Spain," *The Monitor* [Online]. Available: http://www.themonitor.com/news/spanish-king-telling-chavez-to-shut-up-becomes-ringtone-hit/article_43c74a7e-afcb-5aae-8f4f-ff2ec446ec08.html (accessed October 29, 2015).

(The original version of the column appeared in "AP-S turnstile," *IEEE Antennas and Propagation Magazine*, vol. 49, no. 6, pp. 142–143, December 2007.)

NOTE

1. A related topic "Criminal Interference" is discussed in Section 7.8.

DID YOU KNOW?

A FUN QUIZ (IX)

The original version of this quiz appeared in the very first issue of the *IEEE Microwave Magazine* in 2000.

1. If a digital cell phone were made with vacuum tubes instead of transistors, it would be as large as
    (a) A microwave oven
    (b) A minivan
    (c) A British telephone booth
    (d) The Washington Monument

2. Assuming 250 million American users of mobile phones (a 2015 estimate from *Statista*), how many new cases of brain cancer would be expected each year *anyway* among these 250 million people?
    (a) About 150
    (b) About 1500
    (c) About 15,000
    (d) About 150,000

3. The Iridium satellite-telephone network was so named because the original plan was to have 77 satellites (the number of electrons in an atom of iridium). That number was later reduced. To be consistent, which element should the system now be named after?
    (a) Gadolinium
    (b) Dysprosium
    (c) Cesium
    (d) Radium

4. A recent opinion research survey for CNET and Techies.com asked people to choose the century's top "technologists" (voters could choose up to three). Guglielmo Marconi, father of radio, was
    (a) First
    (b) Fifth
    (c) Tenth
    (d) Not included in the top ten

From ER to E.T.: How Electromagnetic Technologies Are Changing Our Lives, First Edition. Rajeev Bansal.
© 2017 by The Institute of Electrical and Electronic Engineers, Inc. Published 2017 by John Wiley & Sons, Inc.

5. Who is credited with introducing the term microwaves (in its modern sense) into the technical literature?
   (a) Marconi
   (b) Schelkunoff
   (c) Varian
   (d) Carrara

6. Which of the following *emits* microwave radiation?
   (a) The human body
   (b) The sun
   (c) Police radar
   (d) All of the above

7. Penzias and Wilson shared a Nobel Prize for their experimental discovery of the Cosmic Microwave Background Radiation at a temperature of
   (a) 2.7 K
   (b) 4.2 K
   (c) 5 K
   (d) None of the above

8. After a mere 40 years of space flight there is already a litter problem in the earth's orbit. How many pieces of space debris are routinely tracked by the American Air Force?
   (a) 850
   (b) 8500
   (c) 17,000
   (d) None of the above

9. Which group performed the song "Microwave Love"?
   (a) The Beatles
   (b) Les Horribles Cernettes
   (c) Absolute Zero
   (d) None of the above

10. Hedy Lamarr (1914–), a movie actress from the golden age of Hollywood (whose likeness appears on packages of CorelDRAW software), is best remembered for
    (a) Making the statement: "Any girl can be glamorous. All she has to do is stand still and look stupid."
    (b) Co-inventing a frequency-hopping radio-controlled system for guiding torpedoes during WWII
    (c) Receiving the 1997 Electronic Frontier Foundation Award
    (d) All of the above

ANSWERS

1. (d) The Washington Monument
Source: *Scientific American*, Special issue on the Solid State Century, January 1998.

2. (c) About 15,000
According to the Center for Devices and Radiological Health (part of FDA), brain cancer occurs in the US population at a rate of about 6 new cases per 100,000 people each year. Therefore, about 15,000 cases of brain cancer would be expected each year among 250 million people *whether or not they used their phones*. A key *unanswered* question is whether the risk of getting a particular form of cancer is greater among people who use mobile phones than among the rest of the population. At the moment, "the available scientific evidence *does not* demonstrate any adverse health effects associated with the use of mobile phones."
Source: FDA's Consumer Update on Mobile Phones, October 20, 1999.

3. (b) Dysprosium (atomic number 66)
Source: *The Economist*, November 7–13, 1998.

4. (b) Fifth
Marconi received 30% of the votes and was led by Bill Gates (48%), Henry Ford (46%), the Wright brothers (41%), and John Mauchly, developer of ENIAC computer (38%).
Source: *USA Today* [Online]. November 9, 1999.

5. (d) Carrara
According to an article by G. Pelosi (*IEEE MTT Newsletter*, Fall 1995), the Italian physicist Nello Carrara was the first one to use the term microwaves (*microonde* in Italian) in a 1932 paper in the first issue of *Alta Frequenza*.

6. (d) All of the above
Yes, even the human body *emits* microwave radiation (as part of the tail of its black-body radiation).
Source: *IEEE Potentials*, August/September 1997.

7. (a) 2.7 K
The 1964 discovery was considered strong experimental support for the Big Bang Theory. Earlier (1949) Gamow et al. had theoretically predicted cosmic background radiation at 5 K (R. Morris, *Cosmic Questions*, Wiley, 1993). Incidentally, 4.2 K is the normal boiling point of liquid helium.

8. (b) 8500
Source: *The Economist*, November 7–13, 1998.

9. (b) Les Horribles Cernettes
According to the *New York Times* (December 29, 1998), the Cernettes, a four-woman amateur singing group (with a changing membership) was based at

the high-energy physics laboratory CERN near Geneva. The words and music were provided by Silvano De Gennaro, a computer scientist at CERN. Just in case you were wondering, here are the opening lines of "Microwave Love" (downloaded from the CERN website):

I am burning inside since you told me

that I am the one that warms up your heart

My love is so wide and when you touch me

I melt and burn in this microwave love.

Anybody ready for an encore?

10. (d) All of the above
Yes, Hedy Lamarr and her Hollywood neighbor George Antheil, an avant-garde composer, were granted US Patent #2,292,387 in 1942 for their frequency-hopping communication system. According to the *American Heritage of Invention & Technology* (Spring 1997), Lamarr (who had been married before the war to Fritz Mandl, a leading Austrian armaments manufacturer) and Antheil came up with the idea in 1940 while playing the piano together, carrying on an improvised musical dialogue up and down the keyboard. In fact, their proposed embodiment used perforated paper rolls similar to player-piano rolls and utilized 88 frequencies, the exact number of keys on a piano. Though the US Navy balked at the idea of putting a "player piano in a torpedo" (in George Antheil's words), their patent has been cited as the seminal work by later patents in the area of (electronic) frequency-hopping systems. Hence the belated recognition in the form of the 1997 Electronic Frontier Foundation Award.

(The original version of the quiz appeared in "Microwave surfing," *IEEE Microwave Magazine*, vol. 1, no. 1, pp. 37–79, March 2000.)

CHAPTER 10

LIFELONG LEARNING

"There is no fruitful science without interdisciplinarity. When you want to be a narrow-minded nerd, leave science and seek employment in a post office!"

—Richard Ernst (1933–)
Recipient of the 1991 Nobel Prize in Chemistry

10.1 BACK TO BASICS

Heisenberg is scheduled to give a lecture at MIT, but he's running late and speeding through Cambridge in his rental car. A cop pulls him over, and says, "Do you have any idea how fast you were going?"

"No," Heisenberg replies brightly, "but I know where I am!"

From ER to E.T.: How Electromagnetic Technologies Are Changing Our Lives, First Edition. Rajeev Bansal.
© 2017 by The Institute of Electrical and Electronic Engineers, Inc. Published 2017 by John Wiley & Sons, Inc.

Michael Rubner, a materials scientist at MIT, who tells the above joke in Reference [1], adds ruefully: "Now, you tell that at a cocktail party, and people will walk away from you. Tell it in front of five hundred eighteen-year-olds at MIT, and they just roar."

It was half a century ago when C. P. Snow, who was both a novelist and a physicist, lectured and wrote [2] about the idea of the "Two Cultures," namely those of the sciences and the humanities. As Tim Adams noted in the *Observer* [3], Snow would sometimes employ a simple test at dinner parties to make his point about the gaping divide between the literary and scientific cultures.

> *"A good many times," he suggested, "I have been present at gatherings of people who, by the standards of the traditional culture, are thought highly educated and who have with considerable gusto been expressing their incredulity at the illiteracy of scientists. Once or twice, I have been provoked and have asked the company how many of them could describe the Second Law of Thermodynamics. The response was cold; it was also negative. Yet I was asking something which is the scientific equivalent of: have you ever read a work of Shakespeare's?"*

The *Observer* decided to do its own informal investigation on Snow's cultural divide by posing a set of basic scientific questions to a celebrity panel of three writers, three scientists, and two broadcasters, with results that would have broadly confirmed Snow's hypothesis. Here is a sampling of the answers [3]:

Q. Why is the sky blue?

Kirsty Wark (BBC broadcaster): Because it is a reflection of the oceans on the planet.... (For the right answer, I recommend taking a look at Reference [4].)

Q. What happens when you turn on a light [switch]?

John. O'Farrell (Writer): I am running out of steam here. I really don't know.

Natalie Angier, a Pulitzer Prize winning science writer for the *New York Times*, offered her own antidote to this new age of scientific ignorance. Her book *The Canon: A Whirligig Tour of the Beautiful Basics of Science* [1] is a spirited attempt in nine short chapters ranging from probability to molecular biology to introduce the fundamental concepts (the "beautiful basics") of science. For example, in discussing charged bodies, she really probes about the meaning of charge. Here's physicist Ramamurti Shankar's take on it as reported by Angier: "A charge is an attitude; it is not in itself anything. It is like saying a person has charisma." And, what about the Second Law of Thermodynamics (about entropy), Snow's litmus test for scientific literacy? According to Angier, "Entropy is like a taxi passing you on a rainy night with its NOT IN SERVICE lights ablaze, or a chair in a museum with a rope draped from arm to arm, or a teenager...The darkest readings of the Second law suggest that even the universe has a morphine drip in its vein."

Angier's rationale for trying to break down the cultural divide: "Of course, you should know about science, for the same reason Dr. Seuss counsels his readers to sing with a Ying or play Ring the Gack: These things are fun, and fun is good." That is good enough for me.

REFERENCES

[1] N. Angier, *The Canon: A Whirligig Tour of the Beautiful Basics of Science*, Houghton Mifflin, 2007.

[2] C. P. Snow, *The Two Cultures*, Cambridge University Press, 1993 (a new edition with an introduction by Stefan Collini).

[3] T. Adams, "The New Age of Ignorance," *The Observer* [Online]. Available: http://www.theguardian.com/science/2007/jul/01/art (accessed December 17, 2015).

[4] R. Bansal, "AP-S turnstile: roses are red, violets are blue...," *IEEE Antennas and Propagation Magazine*, pp. 128–129, August 2005.

(The original version of the column appeared in "AP-S turnstile," *IEEE Antennas and Propagation Magazine*, vol. 49, no. 4, p. 150, August 2007.)

NOTES

1. Reference [4] is included in this book as Section 2.2.
2. For a refresher on the Heisenberg Uncertainty Principle, try for example: http://hyperphysics.phy-astr.gsu.edu/hbase/uncer.html (accessed May 10, 2016).
3. For a more prosaic definition of entropy (and the second law of thermodynamics), see for example: http://hyperphysics.phy-astr.gsu.edu/hbase/therm/entrop.html (accessed May 10, 2016).

10.2 PREACHING TO THE CHOIR?

In 2005, Addison-Wesley published *Feynman Lectures on Physics: The Complete and Definitive Issue*, which includes updates and fixes to the classic physics text from Feynman and other sources. The arrival of this edition was accompanied by the publication of a new supplemental volume *Feynman's Tips on Physics*, containing four lectures, which Feynman delivered on problem solving but which were not included in the original book [1]. In the preface to the 1963 edition, Feynman, who went on to receive the Nobel Prize in Physics in 1965, explains the motivation behind his lecture series:

> *"The special problem we tried to get at with these lectures was to maintain the interest of very enthusiastic and rather smart students coming out of the high schools and into Caltech. … They were made to study* [in traditional courses] *inclined planes, electrostatics, and so forth, and after two years it was quite stultifying. The problem was whether or not we could make a course which would save the more advanced and excited student by maintaining his enthusiasm."*

Feynman's book has been translated into many languages and has been used by physics students all over the world. Some of these physicists recently wrote to *Physics Today* [2] offering glowing words of praise for the Feynman Lectures on Physics. However, back in 1963, Feynman himself felt more dubious about the success of his instructional approach. As he wrote in the preface:

> *"When I look at the way the majority of the students handled the problems on the examinations, I think the system is a failure. Of course,…there were one or two dozen students who—very surprisingly—understood almost everything in all the lectures…. These people have now, I believe, a first-rate background in physics—and they are, after all—the ones I was trying to get at. But then, 'the power of instruction is seldom of much efficacy except in those happy dispositions where it is almost superfluous' (Gibbons)."*

Is the Feynman lecture series then the equivalent of preaching to the choir? Should our best and the brightest physics educators be casting a wider net? Carl Wieman, who shared the 2001 Nobel Prize in Physics for his work on Bose–Einstein condensation, decided to take the "low" road. In an article [3] in *Physics Today*, he posed the question:

> *"How successfully are we educating* **all** [emphasis added] *students in science? This objective is very different from in the past, when the goal of science education was primarily to train only the tiny fraction of the population that would become future scientists."*

In an interview [4] with *The New York Times*, Wiemen, who now teaches at Stanford, noted:

> *"Many science professors aim only to produce more scientists when they teach. They teach to one-tenth of 1 percent of the students. That's not good for society.*

It's producing a citizenry that thinks of science as having no connection to their lives."

Wieman, who (while at the University of Colorado) once taught a basic two-term undergraduate course for nonscientists, The Physics of Everyday Life, used his Nobel Prize money to buy "staff and technology to do things differently in class. A lot of it went to pay people to work with me to create interactive simulations that teach students basic physics concepts." These simulations, which explore topics such as radio waves and microwave ovens, can be downloaded free from the University of Colorado website [5].

Despite their vastly different pedagogical approaches and different target audiences, Feynman and Wieman would have readily agreed on one point: convincing your peers about the need for a transformation of the educational process is never an easy sell. Matthew Sands, one of the co-authors of the Feynman Lectures, once recalled [1] that the initial response to his proposal to revamp the Caltech physics sequence by having Feynman teach the course was not enthusiastic. For example, Robert Leighton, another co-author of the Lectures, initially felt: "That's not a good idea. Feynman has never taught an undergraduate course. He wouldn't know how to speak to freshmen or what they could learn." Wieman spent his 2004 sabbatical writing 29 grant proposals to change the way science is taught. Despite his Nobel Laureate status, all proposals except one were turned down. Fortunately, Wieman did not give up.

REFERENCES

[1] M. Sands, "Capturing the wisdom of Feynman," *Physics Today*, pp. 49–55, April 2005.

[2] *Physics Today* website [Online]. Available: http://scitation.aip.org/content/aip/magazine/physicstoday (accessed December 17, 2015).

[3] C. Wieman and K. Perkins, "Transforming physics education," *Physics Today*, pp. 36–41, November 2005.

[4] C. Dreifus, "Physics Laureate Hopes to Help Students Over the Science Blahs," *The New York Times*, November 1, 2005.

[5] Interactive Simulations for Science and Math [Online]. Available:http://www.colorado.edu/physics/phet/web-pages/index.html (accessed December 17, 2015).

(The original version of the column appeared in "AP-S Turnstile," *IEEE Antennas and Propagation Magazine*, vol. 47, no. 5, p. 150, October 2005.)

NOTE

1. To get a flavor of Feynman's distinctive style, try the light-hearted autobiographical volume: R. Feynman, *Surely, You Are Joking Mr. Feynman!*, Norton, 1997.

10.3 THE OTHER DAVOS?

"There is no fruitful science without interdisciplinarity. When you want to be a narrow-minded nerd, leave science and seek employment in a post office!" [1].

—Richard Ernst
Recipient of the 1991 Nobel Prize in Chemistry

You have probably heard of the annual meetings in Davos, sponsored by the World Economic Forum [2] in January, which bring together the movers and the shakers of the business and the political world with a sprinkling of noted academics to discuss the pressing global problems of the day (climate change, anyone?) in the scenic Swiss Alps. Well, when the weather warms up, not too far from Davos, in the beautiful Bavarian town of Lindau (on Lake Constance), another round of annual meetings takes place, where the intellectual prowess per square meter approaches a delta function. These are the Lindau Nobel Laureate Meetings [3], where past Nobel laureates mingle freely with invited young scientists.

The Lindau Nobel Laureate Meetings were started by two local physicians Franz Karl Hein and Gustav Parade, who were hoping "to help jump start the intellectual recovery" in post-war Germany. The first meeting in June 1951 attracted seven Nobel laureates, who lectured some 400 physicians and researchers. The 2010 meeting, which focused on interdisciplinary research, had "a record attendance of 61 laureates and 650 young researchers from 72 countries." [4]

In conjunction with the 2010 Lindau meeting, the British scientific journal *Nature* set up a website where young scientists (or really anyone curious) could pose questions to the participating laureates. Other visitors to the site were able to vote on these questions. By the time submissions closed, 205 questions had been submitted and over 14,000 votes had been cast [1]. *Nature* selected some of the questions from this website and published them along with their answers by the laureates in a special supplement [5]. To pique your curiosity, here is a small sample of the Q&A:

How can the public be convinced of the importance of fundamental research with no application in sight?

Arno Penzias (Nobel Prize in Physics 1978 for co-discovering the existence of cosmic microwave background radiation) responded: *"...Rather than promoting fundamental research as an abstract concept, I think we do better when we focus upon support for research universities—and the problem-rich environments they create and nourish—as our civilization's most fruitful keys to progress. Examples of the social benefits abound, most visibly by the 'Silicon Valleys' that have sprung up around a number of them."*

What is the one discovery that would herald a scientific revolution in the twenty-first century?

Gerardus 't Hooft (Nobel Prize in Physics 1999 for elucidating the quantum structure of electroweak interactions in atoms) answered: *"I think there will be many, but there is one field that is developing faster than many others: information and communication technology. I think that the discovery of genuine artificial intelligence (computer programs that have learned to think like a human) would have tremendous consequences. Today, many scientists think that it is impossible, but I don't. If such an intelligence can be constructed, it will quickly outsmart humans by a big margin. The consequences are difficult to predict, and even potentially dangerous—I don't fear that such a program will overthrow humanity or anything like that, but it might bring unparalleled power to those who are in possession of such a device."* And this was *before* IBM's Watson demolished its human competition in Jeopardy!

REFERENCES

[1] M. Greyson, "Curiosity aroused," *Nature*, vol. 467, no. 7317, Suppl. S2–S3, 14 October 2010. Available online: http://www.nature.com/nature/journal/v467/n7317_supp/index.html#out (accessed December 18, 2015).

[2] World Economic Forum website [Online]. Available: http://www.weforum.org/ (accessed December 18, 2015).

[3] The Lindau Nobel Laureate Meetings website [Online]. Available: http://www.lindau-nobel.org/ (accessed December 18, 2015).

[4] J. Simmons, "Lindau and the zeitgeist," *Nature*, vol. 467, no. 7317, Suppl. S14–S15, October 14, 2010. Available online: http://www.nature.com/nature/journal/v467/n7317_supp/index.html#out (accessed December 18, 2015).

[5] Nature Outlook: Science masterclass, *Nature*, vol. 467, no. 7317, Suppl. S1–S23, October 14, 2010. Available online: http://www.nature.com/nature/journal/v467/n7317_supp/index.html#out (accessed December 18, 2015).

(The original version of the column appeared in "AP-S turnstile," *IEEE Antennas and Propagation Magazine*, vol. 53, no. 2, p. 148, April 2011.)

NOTE

1. Penzias' wok on the cosmic microwave background radiation is discussed also in Section 2.4.

10.4 MIRROR, MIRROR ON THE WALL; WHO IS THE FAIREST OF THEM ALL?

In 2003, Robert P. Crease, a philosopher and a historian, asked the readers of his column in *PhysicsWorld* [1] to send him their choices for "the greatest equations of all time." He admitted that he was inspired in this quest in part by Graham Farmelo's collection of essays *It Must Be Beautiful: Great Equations of Modern Science* [2], which had been reviewed in a previous issue [3] of *PhysicsWorld*. Farmelo took the epithet "modern" to mean the twentieth century. Therefore, while Fermalo's contributors discuss, among others, equations from the field of relativity (yes, $E = mc^2$ made the list), information theory (Shannon's equation), and SETI (the Drake equation, which gives the expected number of sources of communication from extraterrestrial intelligences), Maxwell's equations (ME) are *not* qualified for inclusion in this collection. (Interestingly, Frank Wilczek, writing on the Dirac equation, manages to sneak in the following tribute to the ME by way of a quote from Heinrich Hertz: "One cannot escape the feeling that these mathematical formulae have an independent existence and an intelligence of their own, that are wiser than we are, wiser even than their discoverers, that we get more out of them than was originally put into them."

Crease's search for the *all-time* greats in the world of science equations has precedents, for example, in the form of Michael Guillen's book *Five Equations That Changed the World: the Power and Poetry of Mathematics* [4]. Guillen's list includes the equations describing Newton's universal law of gravity, Bernoulli's law of hydrodynamic pressure, Einstein's $E = mc^2$ (which has another whole "biography" [5] devoted to it), Claussius's formulation of the second law of thermodynamics and Faraday's law of electromagnetic induction (one of Maxwell's curl equations).

In the October 2004 issue [6] of *PhysicsWorld*, Crease reported on the results of his poll to find the greatest equations of all time. Purists could quibble about the differences between formulas, equations, and identities, but, in discussing his findings, Crease chose to interpret the term broadly. That still left open the question about what qualifies an equation as great, and, as expected, respondents had different criteria in mind while making their choices. If simplicity is the touchstone, it is tough to beat the minimalism of $1 + 1 = 2$, and, indeed, the equation had its share of admirers. Other factors mentioned were the practical utility (e.g., the compound-interest equation) and historical relevance (e.g., the Balmer series for the frequencies of light emitted by the hydrogen atom, which had a century-long history stretching to Bohr's atomic model).

In the final analysis, the two equations getting the most votes were Euler's equation ($e^{i\pi} + 1 = 0$) and (you guessed it) Maxwell's equations. As for Euler's equation, one respondent asked rhetorically [6]: "What could be more mystical than an imaginary number interacting with real numbers to produce nothing?" As for Maxwell's equations, the words [6] of the philosopher Immanuel Kant, although written in a different context, seem a fitting description: "When we discover that two or more heterogeneous empirical laws of nature can be unified under one principle that comprises them both, the discovery does give rise to a noticeable pleasure... even an admiration that does not cease when we have become fairly familiar with its object." I conclude this paean to Maxwell's equations with Feynman's observation [7]: "From a long view of the history of mankind - seen from, say, ten thousand years from now - there can be no doubt that the most significant event of the 19th century will be judged as Maxwell's discovery of the laws of electrodynamics. The American Civil War will pale into provincial insignificance in comparison with this important scientific event of the same decade." And, finally, if you have your own favorite equation and you are willing to share your thoughts about it, drop me a line.

REFERENCES

[1] R. P. Crease, "The Greatest Equations Ever," *PhysicsWorld*, May 2004.

[2] G. Farmelo (ed.), *It Must Be Beautiful: Great Equations of Modern Science*, Granta Books, 2002.

[3] J. C. Taylor, "Book Review- It Must Be Beautiful: Great Equations of Modern Science," *PhysicsWorld*, March 2002.

[4] M. Guillen, *Five Equations That Changed the World: The Power and Poetry of Mathematics*, Hyperion books, 1996.

[5] D. Bodani, $E = mc^2$: *A Biography of the World's Most Famous Equation*, Walker & Company, 2000.

[6] R. P. Crease, "The Greatest Equations Ever," *PhysicsWorld*, October 2004.

[7] R. Feynman, R. Leighton, M. Sands, *The Feynman Lectures on Physics*, Vol. II, Addison-Wesley, 1964.

(The original version of the column appeared in "AP-S turnstile," *IEEE Antennas and Propagation Magazine*, vol. 47, no. 2, pp. 104–105, April 2005.)

NOTES

1. For more on Frank Wilczek and Maxwell's equations, see Section 2.1.
2. Section 10.5 provides another perspective on the "greatest equations."

10.5 EQUATIONS REDUX

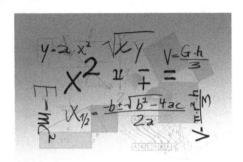

At the end of my April 2005 column [1] on the "greatest equations of all times," I had asked the readers to share their own favorite equations with me. **Allan Love** responded with a copy of the article "The Ten Most Important Equations," which had appeared some 10 years ago in the student magazine *IEEE Potentials* (I don't have the exact citation). The author **D. R. Mack** had asked a group of fellow electrical engineers to list their choices of the top 10 equations that in their opinion had had the most significant impact on our civilization. Mack presciently assumed that Maxwell's equations would appear on everyone's list, so he advised his contributors that they should count the four equations as one to leave room for an additional nine. Based on the input Mack received, he came up with the following subjective list, presented here in rough chronological order (covering a period of approximately 300 years) with my annotations:

1. Newton's Law of Force ($F = m\,a$)
2. Newton's Law of Universal Gravitation
3. The Second Law of Thermodynamics (*corollary*: increasing entropy of the universe)
4. Maxwell's Equations (see Reference [1] for hosannas)
5. The Navier–Stokes Equation (fluid dynamics; see below)
6. The Stefan–Boltzmann Law (black body radiation)
7. Relativity ($e = mc^2$ version)
8. The Lorentz Transformation (changes in mass and time at relativistic speeds)
9. The Schrodinger Wave Equation (foundation of quantum mechanics)
10. Shannon's Sampling Theorem (analog to digital conversion)

Shortly after I received the above list of "earth-shaking" equations from Allan Love, a fellow electrical engineer sent me a provocative quote on one of the equations on the list. The quote comes from *The Physics of Baseball* [2] by Robert Adair of Yale University.

"Almost all of fluid dynamics follows from a differential equation called the Navier-Stokes equation. But this general equation has not, in practice, led to solutions of real problems of any complexity. In this sense, the curve of the baseball is not understood; the Navier-Stokes equation applied to a baseball has not been solved. Professor Emeritus Robert Romer, long-time editor of the American Journal of Physics, *told me of an eminent physicist who said,*

"There are two unsolved problems which interest me deeply. The first is the unified field theory [which describes the basic structure and formation of the universe]; the second is why does a baseball curve? I believe that, in my lifetime, we may solve the first, but I despair of the second."

Well, if you know of an equation that can truly come to grips with the curving baseball, drop me a line.

REFERENCES

[1] R. Bansal, "Mirror, mirror on the wall: who's the fairest of them all," *IEEE Antennas and Propagation Magazine*, vol. 47, no. 2, pp. 104–105, April 2005.

[2] R. Adair, *The Physics of Baseball*, Perennial (HarperCollins), 1990.

(The original version of the column appeared in "AP-S turnstile," *IEEE Antennas and Propagation Magazine*, vol. 47, no. 6, p. 88, December 2005.)

NOTE

1. Reference [1] is included in this book as Section 10.4.

10.6 NEW YEAR'S ~~RESOLUTIONS~~ LAWS

Making a New Year's resolution is easy. But how long will it last? Ah, there is the rub. Instead, as scientists and engineers, we could use that time of the year to reflect on what enduring patterns we have observed in our own *technical* work over the years that could be codified into immutable (well, at least till the next experimental data point dislodges them from their perch) laws. And as a bonus such a law could be named after the person formulating it. A 2003 issue of the *Spectrum* [1] provided some good examples of such laws from the field of electrical and computer engineering, for example, the widely quoted Moore's law, which states that the number of transistors on a chip doubles annually. So give it some thought and if you come up with a nugget of technical wisdom based on your work that you wouldn't be afraid to attach your name to, drop me a line and I would share it with our readers in a future column. To get your creative juices flowing, I have included below some of my favorite *new* laws submitted by natural and social scientists to the online salon *Edge* [2].

Dyson's Law of Artificial Intelligence

Anything simple enough to be understandable will not be complicated enough to behave intelligently, while anything complicated enough to behave intelligently will not be simple enough to understand.

(George Dyson had been exploring the history/prehistory of the digital revolution going back 300 years.)

Barrow's First Law

Any Universe simple enough to be understood is too simple to produce a mind able to understand it.

(John D. Barrow is Research Professor of Mathematical Sciences in the Department of Applied Mathematics and Theoretical Physics at the University of Cambridge.)

McCroduck's Law

A linear projection into the future of any science or technology is like a form of propaganda—often persuasive, almost always wrong.

(Pamela McCorduck is the author or co-author of seven published books.)

Dyson's Law of Obsolescence

If you are writing history and try to keep it up-to-date up to a time T before the present, it will be out-of-date within a time T after the present.

This law applies also to scientific review articles.

(Freeman Dyson is Professor Emeritus of Physics at the Institute for Advanced Study at Princeton.)

Maddox's First Law

Those who scorn the "publish or perish" principle are the most eager to see their own manuscripts published quickly and given wide publicity—and the least willing to see their length reduced.

Maddox's Second Law

Reviewers who are best placed to understand an author's work are the least likely to draw attention to its achievements, but are prolific sources of minor criticism, especially the identification of typos.

(Sir John Maddox (1925–2009) served 23 years as the editor of *Nature*.)

Dawkins's Law of the Conservation of Difficulty

Obscurantism in an academic subject expands to fill the vacuum of its intrinsic simplicity.

(Richard Dawkins is an evolutionary biologist and was the Charles Simonyi Professor for the Understanding of Science at Oxford University.)

REFERENCES

[1] "Commandments," *IEEE Spectrum*, pp. 30–35, December 2003.

[2] The website for the organization 'Edge' [Online]. Available: https://edge.org/contributors/whats-your-law (accessed December 18, 2015).

(The original version of the column appeared in "AP-S turnstile," *IEEE Antennas and Propagation Magazine*, vol. 46, no. 1, pp. 132–133, February 2004.)

NOTE

1. Sections 1.10 and 4.1 are related to Maddox's observations above on "publish or perish" and the ethics of peer review.

10.7 THROUGH A GLASS DARKLY

Kevin Kelly, writing in the magazine *Wired*, reckoned that the internet got properly underway in 1995 with Netscape's wildly successful IPO. In an article [1] celebrating the 10th anniversary of the "birth" of the internet, he decided to rummage through stacks of old magazines to see how the future of the internet looked to the pundits on the eve of the Netscape IPO. What he found was humbling. In late 1994, *Time* magazine rationalized why the internet didn't stand a chance in the mainstream culture: "It was not designed for doing commerce, and it does not gracefully accommodate new arrivals." *Newsweek* was equally dismissive with a February 1995 headline: "THE INTERNET? BAH!"

Kelly's ruminations led Peter Edidin of *The New York Times* [2] to examine the history of some familiar technologies such as radio, television, and film and find out how they appeared to audiences in their early days. Here are some of the brilliant gems he uncovered.

In 1915, the movie mogul D. W. Griffith was interviewed by the *Times* about the future of movies. This is what he saw in his crystal ball: "The time will come, and in less than 10 years, when the children in the public schools will be taught practically everything by moving pictures. Certainly they will never be obliged to read history again."

In 1921, the Russian poet Velimir Khlebnikov imagined in *The Radio of the Future*: "The main radio station, that stranglehold of steel, where clouds of wires cluster like strands of hair, will surely be protected by a sign with a skull and crossbones and the familiar word 'danger,' since the least disruption of radio operations would produce a mental blackout over the entire country, a temporary loss of consciousness."

In 1939, the year of the televised opening of the RCA Pavilion at the World's Fair in New York, the *Times* editors opined: "The problem with television is that people must sit and keep their eyes glued to the screen; the average American family hasn't time for it. Therefore, the showmen are convinced that for this reason, if no other, television will never be a serious competitor of broadcasting."

So what did the future of the internet look like to the experts at the dawn of the twenty-first century? In late 2004, Elon University and the Pew Internet & American Life Project surveyed technology leaders and scholars to forecast the next decade of network development. Their findings included [3]:

- Two-thirds of the experts predict at least one devastating attack on network information infrastructure or the country's power grid in the next 10 years. Some experts believe serious attacks will become a regular part of life.
- Fifty-nine percent of these experts predict increased government and business surveillance as computing devices are embedded in appliances, cars, phones, and even clothing.
- Fifty-seven percent of these experts predict more virtual classes in formal education, with students grouped by interests and skills, rather than by age.
- Fifty-six percent of these experts predict changes in family dynamics and a blurring of the boundaries between work and leisure as telecommuting and home-schooling expand.
- Fifty-four percent look for a new age of creativity in which people use the Internet to collaborate with others and share music, art, and literature.
- Fifty-three percent predict that all video, audio, print, and voice communications will stream to coordinating computers in homes and offices via the Internet.

For the long-term view, we turn again to Kevin Kelly of *Wired* [1]: "**There is only one** time in the history of each planet when its inhabitants first wire up its innumerable parts to make one large Machine. You and I are alive at this moment. Later that Machine may run faster, but there is only one time when it is born…. Three thousand years from now, when keen minds review the past,…this will be recognized as the largest, most complex, and most surprising event on the planet." Savor the moment.

REFERENCES

[1] K. Kelly, "We Are the Web," *Wired*, August 2005.

[2] P. Edidin, "Confounding Machines: How the Future Looked," *The New York Times*, August 28, 2005.

[3] Imagining the Internet. Elon University website [Online]. Available: http://www.elon.edu/e-web/imagining/default.xhtml (accessed December 18, 2015).

(The original version of the column appeared in "Microwave surfing," *IEEE Microwave Magazine*, vol. 48, no. 2, pp. 30–32, April 2006.)

NOTE

1. There is no shortage of books that paint a rosy future for the internet. Here is one that takes a skeptical view:
   A. Keen, *The Internet is Not the Answer*, Grove, 2016.

10.8 STRANGER THAN FICTION?

"I'm gonna live forever

I'm gonna learn how to fly

High"

These high-flying lyrics from the hit musical *Fame*, especially that bit about living forever, may seem like wishful thinking to most of us but futurist Ray Kurzweil would take them seriously. I first heard about Kurzweil and his predictions through Deborah Rudolph, former Manager, Technology Policy Activities for IEEE-USA when she forwarded an *InformationWeek* article [1] based on an interview with him. Futurology is a dicey business and my first instinct is always to check the bona fides of the person making the bold claims.

Ray Kurzweil has been described [2] as "the restless genius" by the *Wall Street Journal*, and "the ultimate thinking machine" by *Forbes Inc.* magazine ranked him #8 among entrepreneurs in the United States, calling him the "rightful heir to Thomas Edison," and *PBS* included Ray as one of 16 "revolutionaries who made America." He was inducted in 2002 into the National Inventors Hall of Fame, established by the US Patent Office. He received the $500,000 Lemelson-MIT Prize as well as the 1999 National Medal of Technology. His website credits him with pioneering developments in character recognition, the first print-to-speech reading machine for the blind, the first CCD flat-bed scanner, the first text-to-speech synthesizer, the first music synthesizer capable of recreating the grand piano and other orchestral instruments, and the first commercially marketed large-vocabulary speech recognition. His book *The Singularity is Near: When Humans Transcend Biology* [3] carries endorsements from Bill Gates ("one of the leading futurists of our time) and the late Marvin Minsky ("one of our leading AI practitioners").

Kurzweil feels that accelerating advances in computing and nanotechnology will have far-reaching consequences within foreseeable future. For example, a $1000 worth of computation in the 2020s will be 1000 times more powerful than the human brain and in 25 years we will have multiplied our computational power by a billion. Nanotechnology will help everyone replace the "human body version 1.0" with a new or rejuvenated one. As Kurzweil puts it [1], "We'll get to the point where we can stop the aging process and stave off death."

Here are some of the other predictions from Kurzweil [1]:

- We will be getting memory backups for *people* by the late 2030s.
- By the 2020s, nanobots (nano-robots) will be circulating in our bloodstreams to make repairs where needed (remember the 1966 movie *Fantastic Voyage* with a miniaturized rescue submarine traversing the body? Coincidentally, Kurzweil has co-authored an unrelated book called *Fantastic Voyage*).
- By the 2020s, human longevity will be increasing at least one year for every year that passes.
- When you run into someone on the street, background information about the person will "flash" somewhere within your vision. (Some of you may recall the graphics accompanying the hero's actions in the film *Stranger Than Fiction*.)
- In what Kurzweil terms the "emerging field of rejuvenation medicine, we'll be able to create new heart cells from your skin cells and introduce them into your system through the bloodstream." Voila!

As a reviewer of Kurzweil's book [3] pointed out [4]: "What's arresting isn't the degree to which Kurzweil's heady and bracing vision fails to convince—given the scope of his projections, that's inevitable—but the degree to which it seems downright plausible."

REFERENCES

[1] S. Gaudin, "Kurzweil: Computers will Enable People to Live Forever," *InformationWeek*, November 21, 2006. Available online: http://www.informationweek.com/story/showArticle.jhtml?articleID=195200003 (accessed December 18, 2015).

[2] A brief biography of Ray Kurzweil is available online at http://www.kurzweiltech.com/rayspeakerbio.html (accessed December 18, 2015).

[3] R. Kurzweil, *The Singularity is Near: When Humans Transcend Biology (paperback)*, Penguin, 2006.

[4] The *Publishers Weekly* review is available online at: http://www.amazon.com/Singularity-Near-Humans-Transcend-Biology/dp/0670033847 (accessed December 18, 2015).

(The original version of the column appeared in "AP-S turnstile," *IEEE Antennas and Propagation Magazine*, vol. 48, no. 6, p. 143, December 2006.)

NOTE

1. Ray Kurzweil's recent book *How to Create a Mind: The Secret of Human Thought Revealed* (Penguin, 2013) discusses how to reverse-engineer the human brain.

10.9 HIGH FREQUENCY EDUCATION: WHAT DO YOU THINK?

"...Education should be the accumulation of understanding, not just an accumulation of facts..." [1]

—David Pozar in the Preface to *Microwave Engineering*

Few readers, whether educators or practicing engineers, in the field of high frequency (RF/Microwaves) techniques would take issue with Pozar's opening gambit. Nevertheless one needs to choose which *facts* should be included in designing a textbook or more broadly a curriculum in high frequency techniques. Ay, there is the rub.

In a 2005 essay [2] in the *IEEE Microwave Magazine*, Prof. Madhu Gupta of San Diego State University took a crack at the issue of RF/Microwaves curriculum design. Following the dictum "Think Globally, act locally," he found that his MS-level curricular choices were influenced by the needs of the local employers in the thriving wireless communications business. Unlike the defense business that dominated the high frequency markets in earlier decades, the extremely cost-conscious, high-volume-oriented wireless industry demanded a skill set emphasizing mixed-signal design, CAD/CAE tools, circuit optimization, and packaging issues. Given the broad range of subjects that need to be mastered, Gupta concluded: "Microwave educators might find it increasingly difficult to maintain a distinct identity and justify a distinct training path for designers in the future."

Gary Breed, the then editorial director of *High Frequency Electronics*, wrote a report [3] on recent trends in high frequency education. He noted that the "balance between general education and specialized training has been accomplished in several ways" and mentioned the following specific points:

- Often the Master's degree provides the specific training sought by employers since "graduate-level research includes the expected 'hot topics' of wireless systems" such as network management, wave propagation for emerging applications, and reconfigurable digital radios.
- Many "engineering departments have added or modified courses within the normal EE program to raise students' awareness of issues involving RF, microwave, and wireless technologies."
- The debate over "specific training versus solid fundamentals" is going away as wireless technologies have matured and the related subject matter has found a spot among the core subjects at many universities.
- Lifelong learning (aka continuing education) in the form of short courses "will present the greatest challenge over the next several years" because of the wide range of subjects from CMOS RFICs to economic analysis of wireless services to be covered.

In an editorial [4] accompanying his report [3], Breed ruminated over his own life experiences to explore the truth behind some clichés about high frequency education. Taking a cue from that editorial, I would like to invite you to share your thoughts and experiences on the following "truisms" in the area of RF/microwave education:

1. A university degree is just a beginning.
2. What a waste! They should delete ... from the high frequency curriculum ASAP.

3. I wish they had taught me how to …

I hope to share some of your responses in a future column.

REFERENCES

[1] D. M. Pozar, *Microwave Engineering*, 3rd ed., Wiley, 2005.
[2] M. S. Gupta, "Educator's corner: curricular implications of trends in RF and microwave industry," *IEEE Microwave Magazine*, vol. 6, pp. 58–70, December 2005.
[3] G. Breed, "Engineering education: embracing wireless and moving beyond," *High Frequency Electronics*, pp. 32–34, August 2006.
[4] G. Breed, "Those cliches about education are often true," *High Frequency Electronics*, pp. 6–7, August 2006.

(The original version of the column appeared in "AP-S turnstile," *IEEE Antennas and Propagation Magazine*, vol. 48, no. 5, p. 135, October 2006.)

AFTERWORD

And, just to wrap things up, here is the final quiz:

Under which of the following circumstances, should you contact the author (Rajeev.Bansal@uconn.edu):

(a) You would like to offer your comments on the book.

(b) You would like to correct errors you found in the book.

(c) You would like to suggest a topic for one of the author's future columns for the *IEEE Antennas and Propagation Magazine* or the *IEEE Microwave Magazine*.

(d) All of the above.

There are no wrong answers; I would welcome your feedback anytime. Thanks!

(May 2016, Storrs)

INDEX

From ER to E.T.: How Electromagnetic Technologies Are Changing Our Lives, First Edition. Rajeev Bansal.
© 2017 by The Institute of Electrical and Electronic Engineers, Inc. Published 2017 by John Wiley & Sons, Inc.

IEEE PRESS SERIES ON ELECTROMAGNETIC WAVE THEORY

Andreas C. Cangellaris, *Series Editor*
University of Illinois, Urbana-Champaign, Illinois